Tectonics: a building for Earth Sciences at Oxford

Wilkinson Eyre Architects

black dog publishing
london uk

The study of the Earth has turned out to be the most important factor in understanding how humans have interacted with the planet and how they will do so in the future. Every aspect of human life can be traced back to some geological source. Everything from the location of a city to the taste of a wine is affected by the rocks beneath us. Today, modern society is critically dependent on mineral extraction and the production of fossil fuels, which are all in limited supply.

This building will provide a platform for expanding the reading and research of the Earth Sciences and allow this subject the room to grow into its increasing role in our future.
Gareth Roberts, Alumnus and Major Donor

The Wolfson Foundation was delighted to contribute toward the cost of Oxford's new Earth Sciences building. The project fits neatly with the Foundation's emphasis on excellence: a striking piece of architecture housing internationally recognised research.

Following detailed conversations, we received an application from the University in Autumn 2009. It was sent for rigorous peer review, considered by our Science and Medicine Panel before the Trustees made a grant early in 2010. The Foundation (as a capital funder) recognises the importance of buildings. It was very clear that the previous inadequate facilities for Earth Sciences were designed for a time when there were fewer staff requiring less in the way of analytical facilities. But the decision was essentially made on the quality and importance of the research to be undertaken in the new facilities: top-class science by an outstanding group of earth scientists.

For success in most areas of science two key features are required: first rate people and first rate laboratories. Earth Sciences at Oxford had the former and now also has the latter—and can surely look to the future with confidence.
Paul Ramsbottom, Chief Executive, The Wolfson Foundation

Contents

John Hood

Foreword

My arrival in Oxford in 2004, to take up my appointment as Vice-Chancellor, coincided with early stirrings for a new building for Earth Sciences. The discipline at Oxford traces its roots to the seventeenth and eighteenth centuries, although academic programmes did not commence until the second decade of the nineteenth century. By the first decade of the twenty-first century the Department of Earth Sciences had long since become recognised as internationally eminent.

During my introductory visit to the department in Michaelmas Term 2004, I was captivated by the range and quality of research activity, the vibrancy of the scholarly community, and the engagement of undergraduate and graduate students in the life of the department. I was also struck by the well-worn state of the premises. They would later be described in a paper to Oxford's Council as "a patchwork of buildings, varying in age between 100 and 15 years, none of which is suited for a modern laboratory-based science".

The department has a straightforward purpose: "enhancing our scientific understanding of this and other planets". With the passage of time the spectrum of its scientific activity had progressively broadened and deepened. More recently, this had included significant investment in the quantitative and analytical side of the discipline. As might be expected, expensive and sensitive equipment for mass spectrometry and related analytical purposes is required for such research. But frustratingly for the staff, the department's aged premises were not suited for renovation to the clean laboratory facilities necessary to house sophisticated scientific tools. Unfortunately this problem was all too common among Oxford science departments in the early years of this century.

Universities in the UK had endured serious funding deprivations throughout 1980s and for much of the decade of the 1990s. Consequently they had severely curtailed their campus investment programmes. By the late 1990s, the government had fortunately begun not only to recognise the fundamental importance of the country's scientific base for its future economic fortunes, but also to do something about it. It had established a programme to finance reinvestment in the fabric and infrastructure of the nation's academic institutions. Specifically, it had introduced a biennial funding stream it labelled SRIF (later SIF) to facilitate the funding of new capital projects.

By the early 2000s, Oxford University's infrastructure investment capacity was further enhanced when the Delegates of Oxford University Press (OUP), a department of the University, began to make one-off capital transfers to the rest of the University. By the time of my 2004 visit, papers describing the Earth Sciences department's real estate dilemma had begun to percolate through the University's committee system. Capital planning within the University was a somewhat fluid concept then, with many projects vying for limited capital funding. So, when I made that inaugural visit, my colleagues' not-so-subtle lobbying for a new building was hardly a surprise.

Owing to the magnitude of the University's capital appetite, the Council of the University sensibly decided that allocations from its SRIF/SIF grants and from the OUP transfer should be further leveraged by contributions from donors and other funding agencies. With more capital at its disposal and with more rigorous capital planning, the University was able to increase the number of projects it could support. The sponsors of those projects were then able to proceed with detailed planning and design, and with their fund-raising endeavours. The strength of Earth Sciences' academic case propelled it to the top of the new project list promoted by its host division, the Mathematical, Physical and Life Sciences Division.

Undertaking new building projects in Oxford does involve degrees of difficulty that are not encountered at most modern research-intensive universities. Time frames tend to be moderated in their own diffuse way! For example, Oxford is an historic city containing an ancient University, so its planning processes are rigorous. They also tend to be robustly contested. Oxford maintains its Athenian democratic traditions, which mean that its committee processes are often complex and lengthy, involving extensive consultation. Fund-raising has its distinctive complexities, also. Enlisting the support of alumni and friends of the University for a new building project is a long-term undertaking. But before that undertaking can even begin, careful and sometimes lengthy negotiations with the colleges of the University are needed, in order to secure the 'right to approach' potential donors who are alumni of the colleges.

The University's Science precinct in South Parks Road was built out during the twentieth century. If anything, its footprint was over-developed. Any new building today means one or more of the existing buildings must be demolished. The old Central Chemistry Laboratory was located on the site that was eventually selected for the new Earth Sciences building. It was nigh derelict and had been rendered surplus to requirements by the construction of the new Chemistry Research Building on the south side of South Parks Road.

Furthermore, the old Central Chemistry Building had been deemed unsuitable for refurbishment for modern scientific purposes. However, its demolition required public consent. A local interest group, possibly motivated more by sentiment than by any redeeming aesthetic features of the building, lodged an application to have it listed. The government's resulting deliberations were prolonged and frustratingly delayed the project's momentum. There were benefits, nonetheless. Most notably, the delay allowed beneficial design enhancements to be undertaken and for further progress with fund-raising.

The firm of Wilkinson Eyre won the design competition for the building. Their proposal was refreshingly creative, belying the firm's lack of design experience with modern scientific buildings. The proposed visual language was sympathetic to the discipline of Earth Sciences and to the building's physical context in South Parks Road. Consistent with the brief, the interior design featured pleasant, light, open spaces, encouraging of formal and informal collegial engagement: a full atrium, open stairways, attractive teaching, office and laboratory spaces, and a magnificent common room with some of the best views over the historic city and its environs to be found anywhere in Oxford.

In the end the necessary government and City consents were granted. In July 2008, fundraising had progressed to the stage where the University Council was ready to approve the project. Demolition of Central Chemistry commenced immediately. Construction proper of the new building was underway by the following February and was completed in early September 2010. The resulting building is certainly true to its promise: a beautiful and functional home for an eminent and confident department; an added draw-card for staff and students alike and a catalyst for renewed scientific creativity and accomplishment. If its gestation was frustrating, its completion and occupancy are just cause for celebration.

The founder of the Fletcher Construction Company, Sir James Fletcher, was reported to have encouraged successive generations of his managers not to construct buildings "they would have to run past". The design of the new Earth Sciences building meets, indeed comfortably surpasses, that standard. Passers-by stop to admire the building. The occupants are enthused by it. May it prove to be a catalyst for another impressive period in the rich history of Earth Sciences at Oxford. Recalling the department's espoused purpose, there remains so much we have yet to understand about this planet, let alone others.

An architectural appreciation/ Hugh Pearman 10

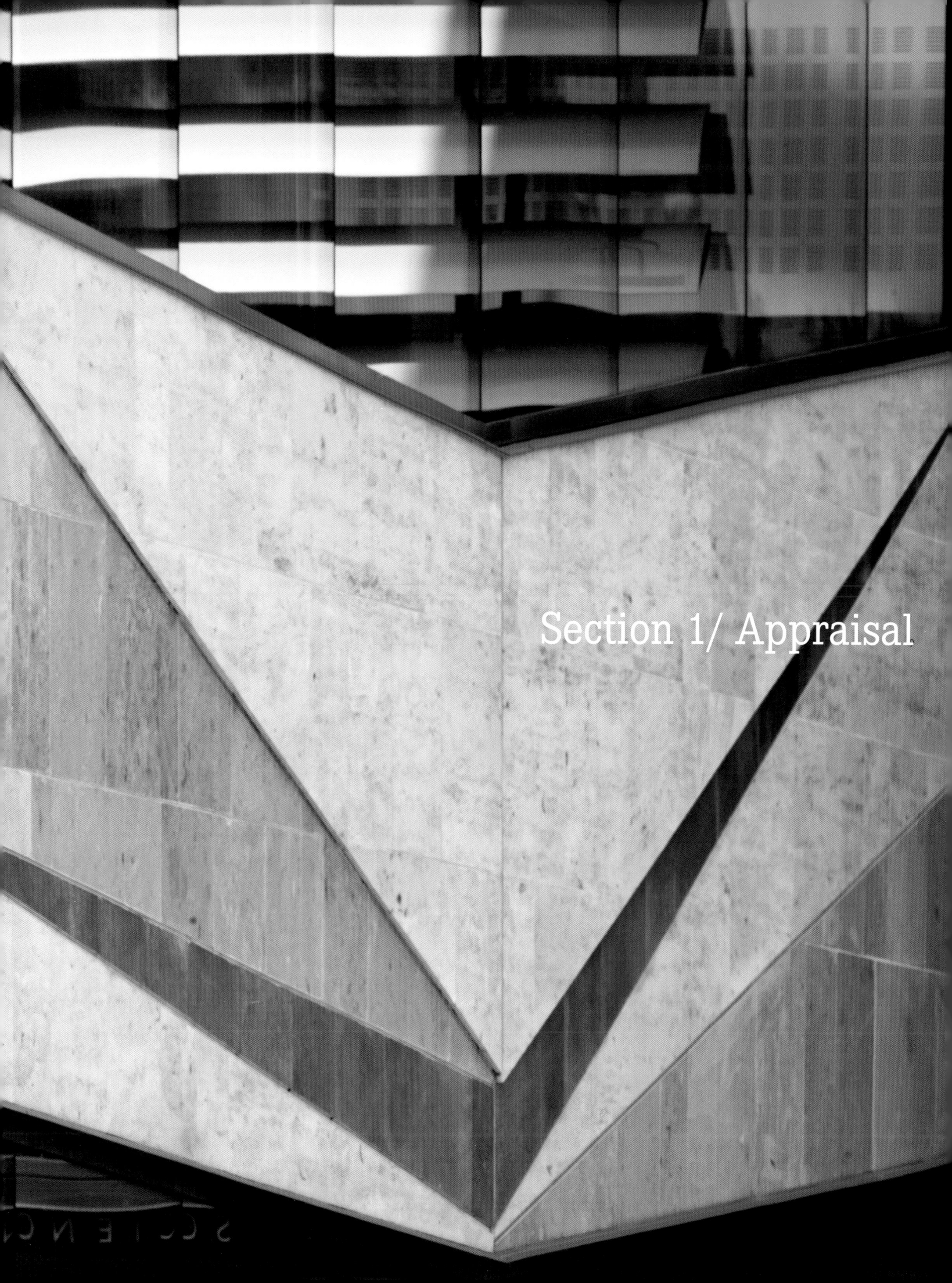

Section 1/ Appraisal

Hugh Pearman

An architectural appreciation

opposite/ Narrative wall, with louvres to atrium reflected.

One's mental image of Oxford–the tourist image, inevitably–is usually of the colleges, seldom of the academic departments. There is a historic reason for this: at one time almost all teaching was done in the hermetic worlds of the colleges themselves, which had relatively little to do with each other. Centrally-based from medieval times, private fiefdoms behind walls and quads, they came to define the city's layout in such a way that even later industrialisation could not usurp them. Pan-University faculty and department buildings arrived centuries later, and transformed the way teaching was done. In particular, the older collegiate universities such as Oxford acquired "science sites" where the rapidly developing newer disciplines could build the laboratory and lecture-room space that the colleges could no longer provide individually. These sites were of course outside the ancient centres, even if they were not very far outside. In Oxford's case, the Science Area lies broadly between the old centre and the University parks, set behind the adjacent High Victorian Natural History and Pitt Rivers museums. A jumble of buildings resulted, accelerating in numbers and scale from the late-nineteenth all through the twentieth and early-twenty-first centuries. It is notable that the central Oxford conservation area ends at the road just before most of the science buildings. This is a different, and increasingly important, Oxford.

The Department of Earth Sciences is part of this concentration of scientific experimentation and learning. Its new building gives it a much more public face than it had previously, and this is appropriate for the increasing importance of what it teaches. Across the spectrum of the sciences, this is the case: research, whether pure or "applied" in the old and increasingly irrelevant distinction, is now perceived more than ever as vital for the economic wellbeing of nations, attracting large grants. The buildings of science departments used to be relatively humdrum compared with those of humanities or the colleges themselves. As a guide to Oxford's architecture disparagingly put it in 1999: "The buildings that make up the Oxford Science Area are mainly twentieth century and are largely of a disappointing architectural calibre." In the twenty-first century, that started to change, and the change is accelerating. Close to the Earth Sciences building, the large Chemistry Research Building (RMJM, 2004) and Biochemistry Building (Hawkins\Brown, 2008) showed the way things were going: big, glossy, self-assertive, affluent-feeling.

Compared to these, Earth Sciences is a somewhat smaller affair. But the same conditions apply, though the process and outcome is rather different. What was once known as Geology is now a discipline which at one extreme has more in common with Physics, using such costly and precise equipment as mass spectrometers, and requiring laboratories isolated from the Earth's magnetic field or from the influence of metals. At the other end of the scale, staff and students still go out and hit rocks with hammers. The fossil record is plainly one important strand of the work here, but Earth Sciences are also economically vital: the world of mineral commodities and earthquake prediction, the behaviour of the oceans and of other planets, for instance, are all part of the picture. So a previously tucked-away department reached the point where it could not only bring together and improve its accommodation and technology, but also give itself a public face.

opposite/ Street view showing articulation of corner of laboratory wing and overhang.
below/ General view from South Parks Road.

There is a lot more to this department than lumps of rock. But be assured that they know their rocks intimately. This is one reason why the cladding of the main wing is Clipsham stone rather than the more usual Bath stone. Both have been used historically in Oxford: but the geologists in the department know that the structure of today's beds of Bath stone is much less dense and more friable than has historically been the case. Clipsham is tougher. Similarly, the department was well placed to advise on the groundwater boreholes that help give the building its low energy consumption. But beyond such pragmatic decisions, and given architecture's enduring love affair with stone, it seems only natural that a geological narrative should have partly driven the appearance of the building. The "narrative wall", a stratified part-relief composition running from the street elevation, along the easterly facade, and on into the atrium of the main wing, brought designers and dons together and is a symbol of the work that goes on here.

The building's plan is straightforward enough: it comprises two linked, slightly splayed wings, one larger and stone-clad with laboratories and lecture theatres and a deep basement level; the other smaller, set back, glass-clad, containing academic and administrative offices (and another pair of ground floor lecture rooms that can be linked when required). The ground floors of both wings are recessed on two sides: slightly in the case of the main wing, considerably in the case of the smaller one. The bulk of the building thus appears to float slightly and provides some cover: in the case of the main block, the angle of the overhang also relates to a slight bend in South Parks Road at this point. Floor-to-ceiling heights are designed to be greater in the laboratory building, the upshot of which is that you get an extra floor for the same height in the office block. The entrance is set well back from the street edge behind what might almost be termed a "quad" in the Oxford manner, one of two in the composition.

below/ The laboratory wing with its rank of chimney flues.
opposite/ View from Le Gros Clark Place.
overleaf/ Night view of the Senior Common Room.

The entrance is into a narrow full-height atrium that forms the hinge-point of the two wings, which share a service core of lifts, stairs and toilets at this point.

The complex is slotted into a tight site, confined by the buildings of various ages and activities around it: the Department of Inorganic Chemistry to the west, the Pitt Rivers Research Centre and Human Anatomy to the north, and the listed L-shaped Dyson Perrins building to the east, originally for chemists (financed by the heir to the Lea & Perrins Worcestershire Sauce business), now mainly for geographers. The second quad, for instance, east of the office wing, is named Le Gros Clark Place after the famous anatomy professor of that name from the department behind. Wilfred Le Gros Clark was one of the team who in 1953 conclusively proved that Piltdown Man was a forgery. It's an indication of the dense and interlocking nature of the Science Area, where external spaces have to serve several functions. Into this complexity the prominent new additions of recent years strut their stuff. Time, perhaps, for a subtler response.

There is also an invisible overhead constraint: a limit imposed by the city planners known as the "Carfax" height restriction after the name of the famous monument in the city centre. Broadly this translates into buildings of four to five storeys, but the planners are aware of the danger of a sliced-off skyline, and so allow certain non-inhabited intrusions to rise through the limit. Oxford has its tradition of spires, and on the science site with its concentration of laboratories this translates into the more prosaic tradition of chimneys and vent shafts. Unlike most of these ad-hoc protrusions, at the Earth Sciences building the design of the chimneys was considered at the outset as a vital part of the overall architecture of the building. The row of four five-metre chimneys, finished in render and metal louvres, are in side elevation akin to an echelon of airliner tailfins or streamlined ship's funnels.

The chimneys are not all that happens up on top of the building, however. While on the laboratory block the necessary plant room is integrated into the form, on the office wing a similar form–this time sans chimneys–is the department's glass-walled Senior Common Room, unusually allowed to break into the Carfax height limit. The necessary setback to reduce the visual impact of this level provides a roof terrace for the dons and their guests. An inhabited rooftop is rare in Oxford, the main other example also being recent: the newly-remodelled city-centre Ashmolean Museum by Rick Mather with its discreet

EXIT

below/ Ground floor undergraduate area, with library beyond.
opposite bottom/ The building unwrapped: concept drawing showing continuity of 'strata' around the facades.

rooftop restaurant. At the Earth Sciences building the effect is salutary: while the lower levels engage visually with the neighbouring buildings, on top the view suddenly becomes very expansive, across Oxford to the surrounding countryside. With a lot of the work of the department taking place in relatively small, cellular laboratories and studies–always the Oxford way–this sense of opening-up, suggested by the client, is a classic architectural release mechanism.

The matter of outlook is addressed also on the ground floor, where an informal meeting and sitting area for undergraduates looks across the landscaped courtyard to the pleasing end elevation of Paul Waterhouse's 1916 Dyson Perrins block (the Waterhouse dynasty, including his father Alfred and his son Michael, also designed Girton College, Cambridge, the first women's residential college in England). This building has come more into focus as a result of the Earth Sciences development, and now has a setting worthy of it. The library, meanwhile, more usually a very private place, here gazes south onto the street. A row of rhomboidal exterior cast-concrete benches, double-functioning as protection from vehicles, fringe the building on the western side of the entrance courtyard.

But to return to the business of Geology and stone: this element of the building's design is what transcends function and neighbourliness, and–while having an ostensible didactic use–serves as the department's visual signature. The competition for this project was under way as Wilkinson Eyre was completing its National Waterfront Museum in Swansea, clad in local slate. They had developed a technique of bonding the stone with precast concrete panels: could this be employed to provide something richer for Earth Sciences? At first there was a thought of somehow using the evocative and colourful detritus of the department's rock-crushing exercises–possibly even behind glass, like the coloured sands of the Isle of Wight's Alum Bay, traditionally sold layered in glass tubes as souvenirs of the Isle of Wight. But it turned out that the quantities involved were too small, and this fanciful notion turned into something more realistic, and visually stronger. The narrative wall is a way of expressing the banded appearance of a horizontal modern building–strip of transparent glazing alternating with strip of solid or opaque spandrel panels. Here the spandrel bands become sedimentary/metamorphic layers of different Jurassic limestones, dominating the facade, with the recessive glazing strips peeping through. The design was abstracted from real geological examples, complete with 'inclusions' of different stone. The task then was to apply an ordering system to natural irregularity, such that the panels could be of a limited number of types. It looks more complex than it is: in fact there are six stone shapes, frequently 'handed', so making 12 variations in all. Glazing comes in a similarly limited number of variations. This was a geometric exercise, if you like an exercise in abstract pattern-making. The stone pieces were then factory-bonded to concrete facade units.

It was a novel exercise for the architects, who are better known for buildings and structures of steel and glass. Even without the narrative wall, this was going to be, on all its other elevations, a punched-window building. In this regard the stone-clad main wing and the glass-clad admin wing are the same concept: it is just that the solid areas of the admin building are made with insulated panels faced in opaque glass. Neither wing has a modernist glass curtain wall. Today's energy-saving standards, warmly embraced by the department which wanted its building

to have a very low carbon output, had much to do with this: in fact the pre-existing building on the site was an energy-hungry 1960s curtain-walled affair.

The narrative wall brings you from the street into the heart of the building. It is a clever technique precisely because it is anything but random. If you unwrap the drawing of the six principal facades of the building and stretch them out into a line, the ordering system becomes clear: straight lines suddenly start to dance along the narrative wall, but the lines do not break. The lowest level is three-dimensional, projecting in relief: on the upper levels it appears to be in relief viewed from certain angles, but this is an illusion created by the relationship of different shades of stone. The paving to the entrance courtyard repeats the striated motif in a more understated fashion, and again runs right into the building on the atrium floor.

The composition of all the facades, which on the smooth elevations clearly express the seams between the panels, is subtly done, as is the punching (or in some cases, projection) of the window units. Games are played with solid and void: the stone facade often appears lighter than the windows, which in daytime are darker. At night this relationship reverses. On the western corner of both blocks, the slightly staggered windows run right round their respective corners, again raising the question of what are the solid blocks, what are the gaps in between?

A wholly transparent full-height glass curtain wall is allowed at one point only: the south-facing entrance facade to the atrium. This brings daylight right into the building but plainly requires shading from full southerly sun: duly provided in the form of angled external aluminium louvres like a giant Venetian blind. The glass of the atrium roof is 'fritted' with ceramic dots to achieve the same end. Daylight also washes into the basement via the time-honoured technique of pavement lights below the narrative wall.

The laboratories in such buildings have to allow for change of use and of sometimes heavy equipment. While the heaviest and most vibration-inducing gear is confined to the basement (along with a mysterious room that removes the Earth's magnetic field entirely) all the laboratory floors can take a lot of weight, ceilings are high, while the windows are removable to lift in heavy bits of kit. Post-tensioning of the floor slabs yielded the necessary strength along with the thinness needed to keep the overall height of the building down. Architecture has to take a back seat in the component-based interiors of such specialist workplaces as the metal-free laboratories–where everything is plastic, so as not to interfere with the readings, and you have to tog up in garb reminiscent of the clean rooms of computer-chip manufacturers. But it has a role to play in other interiors. The professors' studies, for instance, are cellular at the insistence of the department–you will rarely get any university professor, let alone an Oxford one, entertaining the notion of open-plan. But they have glass walls onto the corridors, again bringing daylight in but also making the dons visible. Some are comfortable with this, some less so, and they arrange their rooms accordingly. Larger 'write-up' areas are provided for students. Meanwhile the two interconnected lecture theatres on the ground floor of the same wing have large glass windows looking out across the rear quad with its line of Ginkgo trees set into a lawn. The Ginkgo is a tree as ancient in its lineage as some of the rocks studied in the department, and thus a 'living fossil' akin to the Coelacanth, a fine thing to contemplate, but not necessarily always, especially when projection is required: so the rooms can be blacked out also.

The landscaping strategy has to allow for vehicle entry right past the building to a services yard and boiler house tucked away at the rear. Places to secure up to 300 bicycles were a requirement. And around half of the western facade of the building has to accommodate what you could see as a rather glorious paraphernalia of steaming, frosted liquid gas tanks

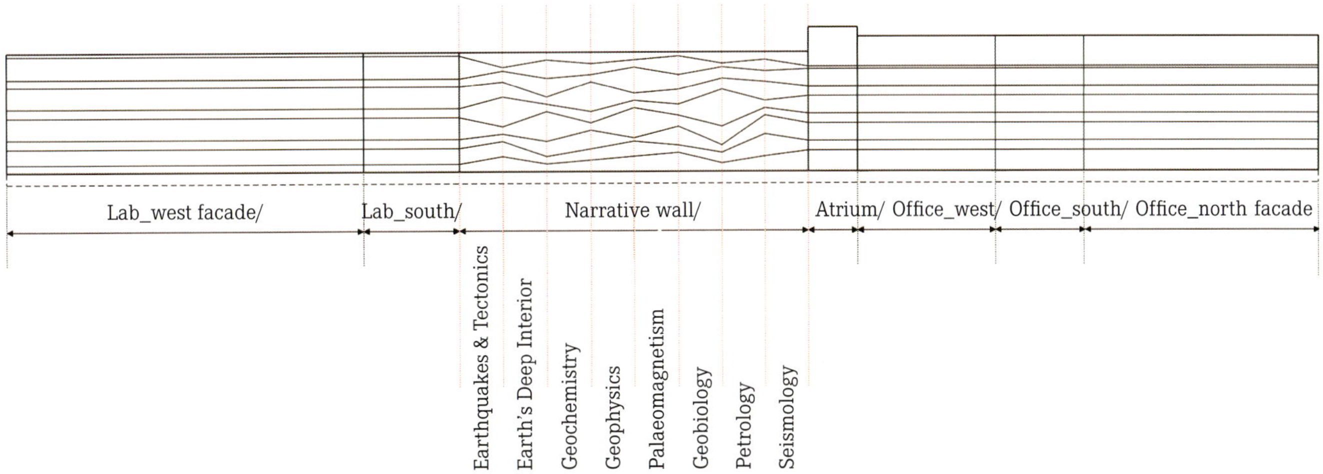

and associated pipework, plus various secure stores and an electricity substation. On the ground floor of this flank, then, are placed the largest lecture theatres which require no great view out. This is the point at which an academic building starts to resemble an industrial one, which is also a reminder of the nature of applied sciences: this is real-world stuff involving real physical and chemical processes. Something of this comes over in parts of the interior. At times it almost resembles a prestige office building–note the designerly ceilings at ground level, which echo the geometry of the narrative wall–at others it is very different and much tougher. For instance the stairs are semi-industrial, though with stone nosings.

This is a building that–rather like the Waterhouse next door–sees no reason for a department to be any less architecturally ambitious than a college. It captures the university character nicely in its contrast of the ultra-concentrated and ultra-private, via a hierarchy of informal and formal meeting spaces, through to the public face. There is conviviality here, along with a clear understanding of what the Department of Earth Sciences is, and does. It has great functional clarity as you would hope: beyond that, it establishes a purposeful new presence in the Oxford streetscape. Most intriguingly, perhaps, it results from a relationship between architects and academics through the shared medium of stone and the Earth it springs from.

opposite/ Night view of narrative wall, showing three-dimensional relief to the lowest layer.
top/ View of atrium from reception area.

left/ Detail of punched windows in office wing.
right/ The office wing from the rear quadrangle.

Section 2/ The department

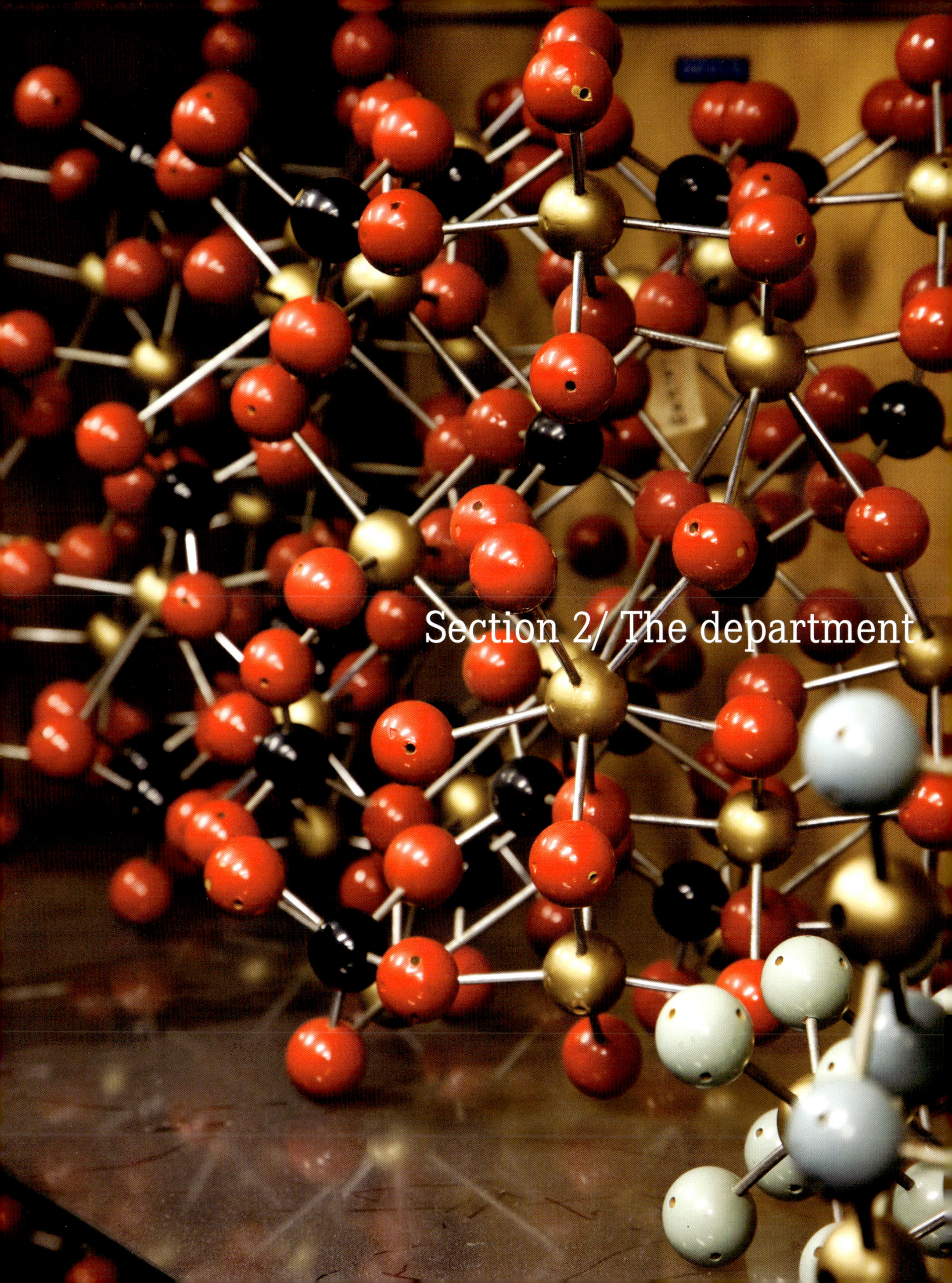

Philip England

The view from here

Geology is the science which investigates the successive changes that have taken place in the organic and inorganic kingdoms of nature; it enquires into the causes of these changes, and the influences which they have exerted in modifying... our planet.
Charles Lyell, *The Principles of Geology*, 1830

At its most fundamental level, the scope of Geology remains as it was when Lyell defined it in the opening words of one of the most influential scientific texts of the nineteenth century. However, like all scientific disciplines, Geology has expanded enormously the range of questions it addresses. In the nineteenth century, and indeed until the 1960s, the geologist's principal tools were the hammer, the hand lens and the microscope, supplemented by some wet chemistry. Now, we use a wide range of physical, chemical and biological techniques to investigate the origins, evolution, present state and–increasingly–the future of our planet. This expansion of scope has been reflected in a change in name of our discipline; I don't think that Lyell would have approved, but Geology is now Earth Sciences.

As far as I can tell, the first Geology lectures in Oxford were given by John Kidd, in the basement of the Ashmolean Museum. Among those who were inspired by Kidd's lectures was William Buckland, who succeeded Kidd as Reader in Mineralogy in 1813. Buckland was one of the great nineteenth century geologists, pioneering the use of detailed scientific analysis to reconstruct environments and events from deep time. Among his achievements were the demonstration that a complex ecosystem must have flourished for a considerable time before the Flood described in the

Bible, the first description of a dinosaur, and the recognition that Britain was covered by ice sheets in the recent past. Buckland was a colourful character–he claimed to have eaten his way through the entire Animal Kingdom (including, perhaps apocryphally, the embalmed heart of a French King)–but he was unable to overcome Oxford's coolness towards the Natural Sciences, and left to become Dean of Westminster, despairing of the University's indifference to the impressive geological collections he had amassed.

Buckland's despair was unwarranted; in 1860 the splendid neo-Gothic University Museum of Natural History opened and, in addition to housing world-class collections, became home to Geology and Mineralogy for the next 90 years. During those years, particularly under the 40-year tenure of William Sollas, the department grew in size until its academic staff numbered 14–a healthy size for a small early-twentieth century science, and one that did not substantially increase until the beginning of the present century. Immediately after the Second World War, the department moved from its cramped quarters in the Museum into the old Clarendon Laboratory, which had been refurbished and enlarged with the aid of a grant from Shell.

In the post-war years, and particularly in the 1960s, Geology boomed; one reason, of course, lay in the subject's pivotal role in the discovery of resources, particularly hydrocarbons. The real transformation, though, came from the massive programme of exploration of the oceans, primarily funded by the US military, and from the expansion of seismology, driven by the need to monitor underground nuclear tests. The combination of these

opposite/ 'Traditional' Geology, and an understanding of the structure of samples from first principles, remains a core subject in the undergraduate course.

below/ Shell's sponsorship of the department's old building, as reflected in the moulding over the door.

two themes led to the discovery of plate tectonics. Although it is commonly stated that plate tectonics caused a revolution in Geology, this statement puts a monochrome wash over what was, in fact, a much more profound and multi-faceted change in the study of the Earth.

It is scarcely a simplification to say that, until the late 1960s, for most geologists, Geology was synonymous with the study of rocks. Of course there were exceptions to this view; Lawrence Wager, Professor of Geology in Oxford until his untimely death in office in 1965, had a clear vision of the path Geology needed to take; he drove, for example, the department's acquisition of its first mass spectrometer in the late 1950s. But such examples were rare, and perhaps it is no coincidence that the UK's contribution to plate tectonics came not from a Geology department but from an idiosyncratic fiefdom on the outskirts of Cambridge, led by a former student of the great physicist Ernest Rutherford.

I am just about old enough to remember those days; I was taught first year Geology in an atmosphere where Mathematics and Physics had no place (except, perhaps, in crystallography) and when, 39 years ago, I wrote an equation on the blackboard during my first seminar in what was then the Oxford Department of Geology and Mineralogy, a senior member of staff walked out, declaring that Mathematics had no place in Geology. Within a very small number of years, however, it was no longer possible to carry out Geology without employing the mathematical, chemical, and physical tools that we now think of as commonplace in the discipline. The real transformation that took place in Geology in the 1960s and 1970s was from an observational science to a physical science.

Oxford kept pace with these developments, steadily adding staff and equipment until, by the middle of the last decade, the department was bulging with a score of mass spectrometers, a hundred computers–including a large parallel machine–and myriad other instruments. With each year that passed, it became increasingly clear that we were unable to carry out modern science in our idiosyncratic agglomeration of buildings–which had been built in five separate phases, dating back to the 1850s.

Nevertheless, we had a lot of affection for our old quarters and, when we came to lay out a brief for the new building–in addition to specifying high-quality laboratory and teaching spaces, we asked the architects to preserve key aspects of our old building–which had shaped the ways in which students, research staff, and faculty interacted over the years. The magnificent researchers' common room at the top of the building provides the forum for many of our discussions, while the serendipitous meetings, which used to occur in the warren of corridors in the old building, now take place in the airy spaces overlooking the atrium.

The new building has enabled us to continue our work across the full breadth of the Earth Sciences, as we now investigate a range of questions that would have been unimaginable to Charles Lyell. Advances in geochemical techniques allow us to probe the origins not only of the Earth, but of the Solar System. The discovery of planets circling distant stars adds a new dimension to Geology:

geochemical data about the birth of our own Solar System now casts light on hundreds of other solar systems.

The changes that are taking place in the climate system pose one of the major challenges facing society. We urgently need to understand the full range of possible climate behaviour so that we can plan for future change. An important part of geological research now is the development of tools for the detection of past climate change, and for very precise dating of events in the recent past. We can now detect annual events, such as El Niños or strong monsoons, in climate records spanning tens of thousands of years.

One of the most fascinating recent developments in Earth Sciences has been the recognition of the jigsaw of interactions between the organisms that precipitate minerals, the chemistry of the oceans and atmosphere, and the Earth's climate. All these interactions leave their isotopic imprint in the rock record. By extracting these chemical signals, geochemists uncover clues about the conditions in ancient oceans and their influence on climate over timescales ranging from hundreds to millions of years. This work is essential to our understanding of how future changes in our climate system will pan out. For centuries, the fundamental mantra of Geology was "the present is the key to the past". Increasingly, we come to recognise that the past is the key to the future.

Until recently, we knew more about the internal structure of the Sun than we did about the interior of the Earth. Seismologists

opposite left/ The Eagle Nebula, a vast cloud of interstellar gas in which new stars are being born.
opposite right/ The study of Earth Sciences is carried out from the largest to the smallest of scales. Thin sections can enable rocks to be studied in forensic detail.

below/ Isotope geochemistry is driven by technological breakthroughs, and the department's strong relationship with mass spectrometry manufacturers such as Nu Instruments has enabled it to develop cutting edge techniques, analysing new elements with higher precision and better resolution than ever before.

use tomography to image the interior of our planet–a CAT-scan of the whole Earth. These data allow us to measure the variations in temperature and composition deep in the Earth, and to understand its inner workings. Seismologists are also preparing to interpret the interiors of other terrestrial planets when seismograms are returned from elsewhere in the Solar System in coming decades. This information will help us to learn whether our world can be viewed as a typical rocky planet orbiting a typical star, or whether a rare, maybe unique, set of events has led to the creation of a rather special planet.

In a parallel approach, high-pressure experiments recreate conditions of the Earth's deep interior in the laboratory. The Earth began as a molten ball formed from interstellar detritus and we do not understand the physical and chemical processes that allowed it to evolve into its present layered structure. How did the core, mantle, and crust separate from one another? How did our fragile envelope of water and atmosphere evolve? Indeed, how did our planet hang on to its water–so essential to life–when others did not?

But Geology is not all about shiny kit. Even after the recent revolutions in techniques, geologists remain natural scientists. Only rarely can we design our experiments; mostly, we pick up and interpret the pieces that are left over after Nature has carried out her own experiments. And this picking up of the pieces takes place in fieldwork–which is the generic label for any programme of measurement or observation made in the natural world, as distinct from the laboratory.

Oxford has field programmes on all the continents, as well as most of the oceans. Sediments of the ocean floor hold detailed records that allow us to document and understand changes in climate over the human time scale, or over millions of years. While the dramatic excursions in Earth's climate revealed by older sediments, now preserved on land, give us insights into likely consequences of humankind's current experiments with the climate system.

For most of the history of Geology, we studied tectonics in order to understand the processes that take place over millions of years within the bowels of mountain chains, or deep beneath volcanoes. We now also use high-precision analytical techniques to examine in high definition processes that take place over a few hundred or a few thousand years. We can extract information from the rocks about physical processes that lead to devastating volcanic ruptions or earthquakes.

None of us can tell where Geology will go in the future. That uncertainty is the affliction, and the joy, of any scientific discipline. It is certain, however, that society has greater need now than at any time in the past for knowledge of the resources and hazards that nature has in store for us; for understanding the present workings and future evolution of our delicate environment, and for the perspective that comes from applying this knowledge to understanding how our own planet sits within the Universe.

opposite left/ The extinction of the dinosaurs was just one of the many times when global changes to the Earth system had a serious impact on life. This band of black shales, examined here in the field by Hugh Jenkyns, records a time, about 90 million years ago, when the oceans became depleted in oxygen and many species died out.
opposite right/ The Arkitsa fault, central Greece, which is the largest exposed fault plane on the surface of the planet. Part of the department's research in tectonics couples field observations with seismology and remote sensing from satellites to improve understanding of the earthquake hazard associated with such faults. Third-year undergraduates on a field trip give the scale.

top/ The caldera of Santorini volcano in the southern Aegean, which erupted about 1620 BC, destroying the Minoan civilisation on the island. This volcano is an active research laboratory for several members of the department. Architect Stafford Critchlow spent time with them on Santorini on a field trip while designing the new building.
bottom/ The cliffs of Santorini provided early inspiration for the narrative wall.

Morley von Sternberg

A snapshot of the old department

Before they moved into their new building, the Department of Earth Sciences invited Morley von Sternberg to photograph their old home. His images document the paraphernalia of 60 years of learning and scientific research. They show the inadequate laboratory and office spaces but the building's eminent suitability for chance encounters, and the souvenirs and objects that express the department's character.

The photographs record the department's many trophies, many of them relating to a rich academic history. The huge antlers belong to a now-extinct Irish elk, and also appear in the lithograph of one of the earliest Oxford Geology lectures hanging on the wall below. A bust of George Bellas Greenough stands sentry to his beautiful–and controversial–1839 geological map of England and Wales, while a spectacular plesiosaur fossil swims above the drinks dispensers. Other artefacts relate to in-jokes and the everyday life of the department. A cupboard is papered with irreverently photoshopped images of the faculty members and ranks of passport photographs contrast with the formal departmental portraits lining the narrow stair. A photograph of Sir Henry Miers, alumnus and mineralogist, is displayed beneath the "Stepfather of Geology" Henry M Odbert. Mr Odbert's self-created portrait hung unnoticed on the wall of a meeting room for some time before anyone realised he was actually an undergraduate in the department, the practical joke gaining him honorary membership of the Sollas, a club for geologists at University College. Elsewhere, the images show the aesthetic qualities of the earth scientist's tools–rows of wrapped-up microscopes and a scattering of laboratory clogs.

EXIT

GEOLOGY AND MINERALOGY

ALASKA
Happy Hour
HEEBEE GEEBEES
ONE NIGHT ONLY
Authoritarian
Libertarian
Fire door Keep shut
STOP PLATE TECTONICS
SHARP EDGES
B4 Layout
Pub Stops of Oxford
Getting closer to earthquakes
Gravity.
It isn't just a good idea.
It's the law.

CLARA

THIRD YEAR
ESSAYS
DEADLINE: 2
APRIL
TWO COPIES
REQUIRED

HENRY M. ODBERT
STEP-FATHER OF GEOLOGY
(2001 ~ 2005)

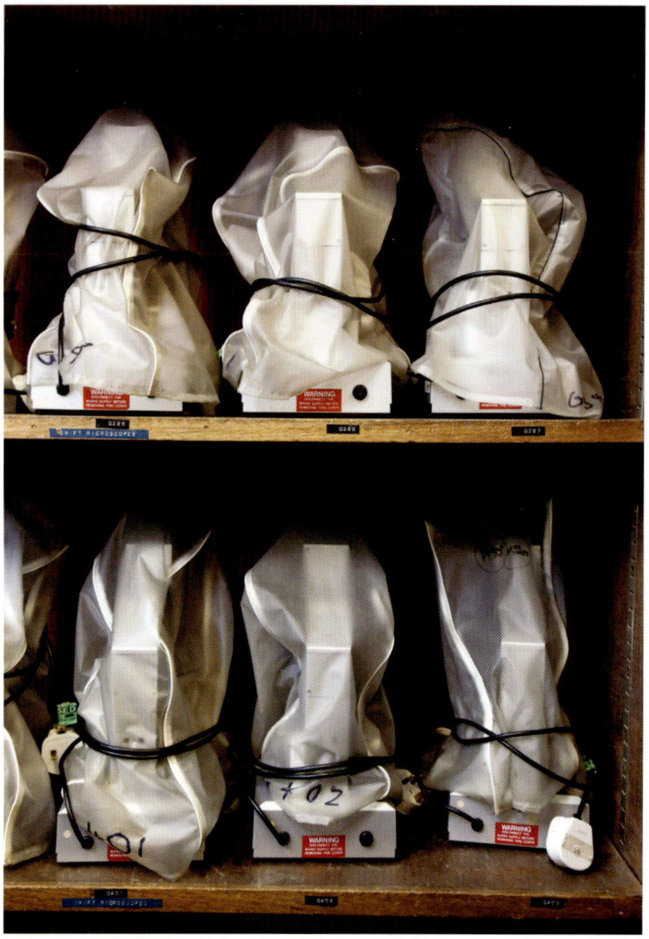
WARNING
WARNING

50p

Everyone's an earth scientist

History

I'm probably the oldest person still coming to this building every day. I first came to the department in 1951 when I was a student, so I've grown up through all the old buildings.

The department is unrecognisable compared to what it was. The Earth Sciences came of age in the 1960s when continental drift and plate tectonics and the modern phase of the subject came in. That was when the whole layout of the building changed, and the design of the labs.

In my early days here, the subject was called Geology. It wasn't taught in schools like other scientific disciplines. It was taught as a series of separate subjects, without too much integration. And in the 1950s and 1960s the exciting thing that happened to the Earth Sciences was that everybody began to realise that this was the study of the origin, history, structure, evolution and age of the Earth as a whole. Everything that's ever happened to it, and therefore rather than learning it as a series of watertight subjects, learning a thousand fossils and their Latin names, it became a study which used Physics, and Chemistry, and Biology, and Mathematics. All the processes that had ever affected the Earth, from the beginning to the present time. And that is what changed the whole subject, the way it was taught and who taught it.

When I first came to the department, continental drift, which is now absolute standard, was still widely disbelieved. And then in the 1960s proof came along. That made an immense difference to the teaching. The whole thing became more integrated. So

now you have this beautiful building and the whole thing is geared so that the students, when they come up to Oxford, they may not have done Geology at school but at least they've done Physics, Chemistry, Biology and Mathematics, then, when they come here, they can be taught the first principles of the Earth Sciences, which is the study of the Earth.

The revolution has arrived in Earth Sciences. What you see here with the Geochemistry labs, the Physics labs, the isotope labs: there will be more of that, and the equipment will improve. But the revolution has occurred already, and for a long time yet there will be a big increase in the technical and analytical facilities. If and when rocks eventually arrive from Mars and the other planets, I can imagine there will be another huge increase in the kind of work being done. Rocks are still the geologist's bread and butter. They are what the Earth is made of. Everything's a rock.
Stephen Moorbath, Professor Emeritus

The department

There's a certain personality to the department because academics are a very quirky breed. And we've brought that to the new building. One of the general features of an Earth Sciences department, and one of the things that we pride ourselves on, is that we're not as rigidly formal as a lot of the other departments. And so maybe more of that personality does come across.

You have people who do a lot of work in the field with students, they're very practical. Students take on a critical role in their own training. They're not passive recipients, they're actively involved. They're more like apprentices to become earth scientists than students. We take that very seriously and as a consequence there's much more open dialogue between members of the faculty and the students. I think that's a general feature of Earth Sciences departments, but it's particularly true in Oxford. It's exaggerated here because of the tutorial system. Students will come in and out of faculty members' offices to do tutorials and so they'll actually be interacting with people who are leaders in their field.
Matt Friedman, Lecturer in Palaeobiology

I certainly see my department as a base from which I go and explore. I go to the field and do fieldwork, and this is somewhere that I come back to. I bring samples from the rest of the world and come back here to work on them. It's like a camp I suppose.
Gideon Henderson, Professor of Earth Sciences

I think we're more of a community than many other departments are. Sometimes when you go to an American department you find that everyone's working at home on their computers and you never see them. You don't get a sense of community, whereas here you do. The fact that the doors have windows means that you get a sense of everyone being around. It is a special attribute of this department that it feels like it's a community. Certainly for students and post-docs, who tend to work very long hours in the labs, it does become a second home.
John Woodhouse, Professor of Geophysics and Head of Department

It's a nice department to work in, quite old fashioned. We all have morning coffee at eleven o'clock. I've been told that three per cent of all research science is evolved over coffee. If you get scientists from different fields every morning at eleven for coffee, they cross-pollinate. That kind of communication really does work.
Toby Christensen, Building Manager

First plans

The original plan was to put new labs into the old building. But when we began to look at getting a new building instead it was a very exciting prospect. If you can start from scratch you can design something really ideal–it's a chance which most people don't have. So it's been a wonderful opportunity to craft the best lab in the world in a brand new building, and we have been able to think very carefully about the environment we have created for the next few decades of research.
Alex Halliday, Professor of Geochemistry and Head of the Mathematical, Physical and Life Sciences Division

Old and new

The issue with the old building was that it was completely unstructured. It came about as a series of accidents, but no-one had given any consideration as to how the spaces would be connected. It's more coherent here.

I don't think the new building has given me more scope for my research, but it's certainly a more organised and structured space. In the old building, my lab ran over three different rooms –it was historical the way that it evolved. Now I can actually organise things much better. I can think about what goes where, and how, and that will improve the workflow. You can get the same research outcomes more quickly, and you can do things in a more organised fashion.
Conall Mac Niocaill, Departmental Lecturer in Earth Sciences

Everything was shoehorned into the old building, so space and facilities were restricted in many ways. When I arrived in 1995 the only geochemist in the building was Steve Moorbath, and the instrumentation was quite long in the tooth. My research group arrived with some instruments and we had a useful amount of space in the newer part of the old building. But as Geochemistry grew, the space very soon ran out and everything got a bit cramped. To get a large state of the art instrument like the 1700 [mass spectrometer] in was just not possible. So the new building has certainly freed everything up. And to get the 1700 –well, there are only four or five in the world, so it puts us in a very strong position. And given the extra space we have, I think we can take on more development projects because we have more flexibility now.

I haven't experienced other departments in terms of the way that the undergraduates relate to the researchers. But here part of the undergraduate course involves getting onto instrumentation, so it's all there for them. And it must make a better impression on students now we have the new building.

The building makes you feel better, more positive. Certainly when you have an environment which is properly structured it's better. In the old department you had to work much harder to get in the right state of mind. But the new laboratory suites in this building make it much easier. People can be much more organised. The nature of the business of research, instilling into people appropriate practice in laboratories, is easier if they don't come from a grubby corridor into a cramped laboratory space. Now we have laboratories opening off a wide corridor. It's infinitely better being here.
Nick Belshaw, University Research Associate

When I started in Earth Sciences I was in the other building. I had a smallish lab and really we were only able to do what I'd started out my career doing. So I could do Geochemistry, I could separate sediments and so on. I could do all the things I'd been trained to do. But then I started to evolve my research in a much more interdisciplinary way. I was evolving this biological arm to my research programme but there was no room at the old building for the lab space I needed, so I ended up with a series of spaces in Zoology. For the last year or so, we were running between different departments trying to keep the research going. The great advantage of the new building for me is enabling this interdisciplinarity to take place under one roof. Suddenly my research group, which had been rather fragmented, has been brought into one building and we can all talk and be together. It's much more cohesive group now.
Ros Rickaby, University Lecturer in Biogeochemistry

I came here in September 2009 so I knew the old building briefly. I had an office with wonderful 1970s curtains, splendid things! But it's been a wonderful shift to come here from a building that was cobbled together to reflect the evolving needs of the department over time. There were other members of faculty that I would almost never see. You'd maybe see them at coffee, but otherwise you wouldn't bump into them in the corridors because effectively you had three separate buildings. Then of course there were additional facilities specifically for Palaeobiology that were across the street. So it's a wonderfully refreshing experience to go from that into a single setting where there's more coherence to everything.

It's not as if I would have turned down the job offer had there not been a new building, but it was certainly an excellent motivation. It became a more attractive offer knowing that I'd be moving into a new facility, not only for my own benefit, but also from the point of view of someone who as a young member of faculty is looking to attract a research lab. There's a world of difference between inviting young postgraduate candidates to come into this bright, open, airy, impressive space and what they would have dealt with before–a cramped undergraduate lounge with a couple of vending machines and a tatty old plaster plesiosaurus on the wall. It's a very different proposition. And that, I think, makes a huge impression on people who come here as candidates.
Matt Friedman, Lecturer in Palaeobiology

Light

I think the light is one of the things that's different, compared to the old building. Oxford has pretty buildings but it's all relatively cramped and small. But if you come in here there's a lot of light. That's the biggest difference between this and the old building, just the amount of space and light. A feeling for the outside world is what attracts a lot of people to Geology, and in that regard the building represents us. A lot of scientists see the work as much more than just a job. People spend nights in the lab, weekends in the office and summers in the field. If you're in a lab with all the machinery running, working 16 hours a day with artificial light, it's not a very pleasant work environment. So it makes a huge difference to have these big windows and all the natural light, it really does. You're less likely to make mistakes.

Even compared to when I was studying in the 1990s the subject has changed. Even though I still work with fossils, with rocks and the physical aspects, it's more computer based and chemistry based now. I think the lab wing reflects that. The change from having a couple of rooms with a machine in them, and huge collections of sample material, it's now the other way round. It's moved from more observation based work to more analytical and quantitative work. But that means you need infrastructure, and that's reflected in the new building. It's a purpose built infrastructure for that kind of change.

We're already using the mass spectrometers in here. And the building being purpose built makes a huge difference because of the air handling and extraction. It's now specifically built for the machines and it's separate from the clean lab. In the old building if the clean lab had a problem, the machines did too because they ran on the same air handling system, which also wasn't up to spec. That caused all kinds of stability issues.
Chris Siebert, Research Fellow in Mass Spectrometry

I'm quite old fashioned so most modern buildings do leave me cold. I'm from the Prince Charles school of architecture. When we were told about the new building we were shown three designs, and this is the one I preferred. The lab wing is the more striking and appealing–the stratigraphy effect is very noticeable. The office side is a bit plain. But the idea of hiding the fume cupboard ducts in the chimneys on the roof is good.

I'm quite fortunate because I've got an office with a window in it here. I've never had that in my career before. My offices have always been rooms within rooms, so I couldn't tell what time of day it was most of the time. It's quite nice to have, after 35 years here, a room with natural light.
Steve Wyatt, Technician, Geochemical Lab

Functional spaces

The metal free labs are interesting. Mike Morris [Laing O'Rourke] says it was the most challenging thing he'd ever done in his life. It was a challenging thing to put together, and it was good to involve external companies too so we got something quite

innovative. I think we're all pretty happy with it. It's a very nice facility that's going to enable us to do a lot of good stuff. And it has brought together different groups across the department. Philosophically, what we wanted to do was have a big facility that could be run as a single lab so that we could bring groups in and they could grow and shrink and interact. We wanted to get a more organic feeling.
Gideon Henderson, Professor of Earth Sciences

We've now got ten metal free labs, so there's bags of room for everybody. Before we were very cramped in comparison. I've got no worries now about going in and doing lab work because there will always be a spare hood. There are no logistical bottlenecks in finding space to work.
Paul Savage, postgraduate student

It's early days in the new building but the physical environment's great. I'm really happy with my office–it's a much more functional environment. It's working well for the different uses I need to put it to–for having research meetings there, for teaching in small groups, and then also for doing my research. And our lab space has improved, but we're still settling in. It's great having all the teaching space in one area–that works really well, having that teaching hub. In a way, the old building was such a hotchpotch that we were forced to bump into each other a lot because it was so inefficient to move around it. The environment's very much nicer to work in here–having everybody under one roof, all mixed up, is good. I think that's really a big advantage.
Tamsin Mather, RCUK Research Fellow

The really important thing that's been done is that the 24 hour access has been preserved. You can't overestimate how much of a benefit that is to students at Oxford. We should be proud of it, the fact that we give responsibility to our undergraduates so that they can come in and have access to the library or the Mineralogy labs at any time they need it. Obviously it's been made easier by the design of the building, and the fact that we've got all the spaces together.
Jack Matthews, undergraduate student

The thing that people have when they walk into Earth Sciences, that you don't really get in other departments, is this sense of the department being their home rather than somewhere else in Oxford. I think it's unusual even among the science subjects. One thing that we all said right at the beginning of the design phase is that we definitely want our own coffee room and we definitely want a student coffee room. They serve two different roles but they're both really important in getting people to mix up. We had to fight the University to keep that space but I'm really glad we did it. It's really important.
Gideon Henderson, Professor of Earth Sciences

If you walk down South Parks Road, and you see the Radcliffe Science Library, and you see the building next door, and then you see this one, the stone's almost exactly the same colour, although they're obviously built in totally different eras. There's a theme there and certainly as a geologist you pick up on it. That puts you in a good frame of mind. You know it's been done with a feeling for what was there before: limestone. You get plenty of light off it.
Hugh Jenkyns, University Lecturer

Aesthetic

It's quite nice to be in a swanky building. You go into college and people have noticed, especially the geographers who are next door. It's a striking building, there's no question about it. I think the narrative wall works brilliantly, it really is very good.
Barry Parsons, Professor of Geodesy and Geophysics

One of the things I do like about this building is the fact that the design of it reflects the subject matter. The theme of sedimentary layers, I like that aspect a lot.
Emma Brown, Academic Administrator

In German we call it a Blickfang–and the building certainly attracts attention. Just from the side where I have my office you can see them, people actually stopping to look at the building or photograph it, and then moving on. It's something that attracts the eye.
Chris Siebert, Research Fellow in Mass Spectrometry

The general response to the building is positive within the Science Area. I'm sure it's done a great deal to raise the profile of Earth Sciences. Before we were by the museum, and I felt more associated with that than anything else. I never saw any of the other departments. And now it feels like we're in the heart of the Science Area.
Richard Walters, postgraduate student

I really like the new building, but then I like this kind of architecture. The simplicity of the large spaces with optically flat white walls is great. It's certainly improved my year. I wouldn't say I would let architecture steer where I work, but it does mean something to me and I do think about it.
William Nash, undergraduate student

I like this building for its beauty and its attractiveness. And its function. The building feels so integrated. When you get to my age and you retire you usually go downhill. But this is going uphill. The design, especially on that side, with the limestones, it's fantastic, really special.
Stephen Moorbath, Professor Emeritus

Integration

We see more of each other in this building than we did in the old building. We see each other walking down corridors and on the stairs, and in the coffee room. It works very well. It is

people's experience that casual chats lead to new ways of thinking about a problem, and new avenues of research.

It was one of our concerns right at the beginning of this project that we didn't want a regimented building where we would all be locked away in different cubicles, and I find that I bump into people more often in this building than I did in the old one. I feel actually more integrated as part of a family of academics here than I did in the old building. That's one of the things we were afraid we would lose.

Morale is quite good, and the new building has only improved it. I think people are very positive about their work.
John Woodhouse, Professor of Geophysics and Head of Department

Process and management

The decant was hard work. Everyone was worried about whether they could do their research and their science. But they were able to do it the day after they moved in.

I was fully involved in the commissioning of the building. You can build a building, but the true sign of a good team is after practical completion, about how they respond when there's a problem. But the contractors haven't run away from anything. There hasn't been anything they couldn't sort out.

I'll be involved in measuring the building's energy use. The building uses a lot of electricity compared to the old one. But that's because we've got a lot more equipment here. The air handling's more sophisticated and we've got more of it. But we could save a lot of money by turning certain bits of kit off at night. I've got all the controls I need to save energy, but we need to find a compromise between the effect that fine tuning will have on the science and on the budget.

The heating and cooling within the building is very efficient–the gas bill is on the floor. We're using hardly any, which is great. The ground source is working but the building's also working in the way that it's designed. It retains the heat and coolth.

I look at this building as mine, and I treat it as mine, because then I care for it. You start to love the building. I love the building. It works, and it works well. Everyone enjoys it, and now we're finishing the landscaping everyone looks at it. That wall, lit up, is unbelievable. It's very dramatic. We've just got to keep changing the light bulbs now.
Toby Christensen, Building Manager

I've been with the department for eight years, so I've been here right through the whole process from the first presentations. This is our department. What we have here is what we asked for, and it is changing the way we work. It has changed certain dynamics, but I think the move has been wholly beneficial to the way we work.
Gill Halstead, Administrator

I think the building is outstanding. I felt the process we went through in terms of making sure it was put up the way we thought it should be put up was very positive. Basically from the architectural end of things we couldn't have asked for better partners. And the construction people were great too. It's gone fantastically well. The project came in on time and under budget. There are always teething problems but we've actually had relatively few of them.

The effect the building is having on people is very important. Not only are the sceptics who weren't really convinced before now convinced, but for everybody there's a sense of it being fun to be here in the morning. People want to come into work. It's such an important thing. Academics don't get paid much and, given how important and how clever they are, how much they contribute and how hard they work, you really want to create an environment where they feel appreciated. It's important in the sciences in particular. What the building contributes is the provision of a place where people can really thrive–we've got no excuse for failure now.
Alex Halliday, Professor of Geochemistry and Head of the MPLS Division

The subject

We're branded Earth Sciences. We used to be called Geology, and of course Geology is from the original Greek for 'the study of the Earth'. And that encompasses everything that we do. And I suppose everybody has some interest in some aspect of the Earth because that's the planet we live on. As far as we know it's the only one that's inhabited. And so it's really important that we understand what's going on on our home planet, because inevitably our future is going to depend on our understanding of the place in which we live. Everybody is to some extent an earth scientist because everybody to some extent tries to understand their home environment.
Conall Mac Niocaill, Departmental Lecturer in Earth Sciences

Earth Sciences is important by virtue of its spread, because it integrates so many different aspects of science, and because so many of the issues that actually face us today are ones that require people who are multivariant thinkers. It's clear when you look at the research profile of the people in the department that you have people who effectively trained as mathematicians and physicists. At one end you've got people who are chemists and at the other end you've got people like me, who are biologists. You mix them together in a place like this and you really get a better understanding of how the natural world operates because you do need to think in all those different disciplines to get a handle on patterns of environmental change, for example, and the biodiversity crisis, and changes in long term climate. I think the perspectives provided by Earth Sciences are really important, because they integrate those different fields. But I think the other important factor is that the Earth Sciences necessarily have a deep time perspective. We think about these issues in a historical context. We have a framework, we think about past changes on Earth and how those relate to modern changes. Because of course the past is the only means we have

to calibrate changes that are occurring now. You could cite that aphorism about how those who forget the past are doomed to repeat it. But having that long term view, that window into deep time, is enormously important I think.
Matt Friedman, Lecturer in Palaeobiology

If you want to know how the world works as an integrated system, you have to take aspects of many sciences and link them together to build a bigger picture, looking at deep space and deep time. Earth scientists work at many scales and within many timeframes. So we'll be looking at tiny grains of sand, but also plate tectonics, and beyond that whole planetary systems. We don't just look at one thing out of context, instead everything is nested inside the greater context of the Earth system.
Richard Walters, postgraduate student

One thing about Earth Sciences is that it helps you appreciate that we need to think about the longer term; you can't just think in a five year cycle, as many politicians do.
Ros Armytage, postgraduate student

If you don't invest in fundamental and applied research it's not clear what your new economy's going to look like. And the UK more than most countries has tended to move away from having a lot of government investment in science. Science has been sold off to private enterprise, and a lot of know-how has been lost. They've let the universities be free agents. But the difficulty is, unless you have a strong science base you run the risk that the whole country could end up in difficulty. The big pharmaceuticals are selling off their science and research know-how too, and at the end of the day they are beginning to realise that they don't have the expertise in-house any more. They don't even know what questions to ask, and what expertise they should be looking for. They don't even know who else they've got within their organisations who might be able to help. You see something similar in government—there's not a thorough understanding of what skill sets are available, what questions to ask and who to ask. And that means it's quite a precarious situation for the UK. It means the universities have to work quite hard to help the UK economy grow in the future.
Alex Halliday, Professor of Geochemistry and Head of the MPLS Division

The above extracts are notes from conversations between Stafford Critchlow, Emma Keyte and members of the department on 2, 9 and 13 March, 2011.

Stephen Hesselbo

An essay in stone

Oxford stone

Ask almost anyone to think of Oxford and their mind's eye will no doubt conjure a picture which includes as its dominant motif the honey hue of the stone from which the buildings are constructed. This is as it should be. On the other hand, ask almost anyone to name that stone and they will invariably be stumped for an accurate answer; in my experience a very tentative "Sandstone?" is the usual response. In fact Oxford has been built over a period of about one thousand years from a variety of different limestones, sedimentary rock composed predominantly of calcium carbonate. These limestones have been sourced at ever increasing distances from the city boundaries. The new Earth Sciences building fits snugly in this evolutionary sequence, but perhaps occupying one ring out from most of its predecessors.

The use of stone in Oxford buildings has been elegantly documented in *Oxford Stone*, a seminal work completed by the great Jurassic geologist WJ Arkell in 1946. Arkell had an agenda, well expressed in his preface which dedicated the book "to the better appreciation and enjoyment of Oxford buildings" and in which he expressed the "hope that it may be of use to those who carry or share responsibility for the maintenance of the buildings, and strengthen those who have to fight against inertia, opportunism, or short-term economy". Arkell foresaw an age when the traditional uses of stone would be either forgotten or willfully ignored. Although construction of the old Earth Sciences building on Parks Road was started in stone in the same year as Arkell's book was published, the post-war years of austerity and utilitarianism

saw very little new building in stone, and constituted a period particularly unfruitful for either traditional or modern architecture in Oxford.

The early stone builders of Oxford, the Saxon inhabitants of the embryonic city, used stone from the hills immediately surrounding the city. The buildings that survive from that time include the tower of St Michael's church on Cornmarket, which was built of rough blocks of Coral Rag; other structures of comparable age that were built from similar materials have survived equally well, including the old city walls and the Castle tower. Coral Rag is, as its name suggests, made up from the hard-cemented remains of coral reefs that grew abundantly in the blue-water seaway that was present in this region during that stage of the Jurassic now known internationally as the Oxfordian, about 160 million years ago. To survive the battering by waves the reefs were constructed for strength, and they have remained strong right to this day. In addition to these hardened blocks, the early builders of Oxford also used so-called freestone, a relatively soft limestone that could be easily cut to shape using a saw, which was used for cornerstones, door jambs, lintels, and sills. All the freestones are "grainstones", rocks composed mostly of cemented carbonate sand, such as is forming from "oöids" on present day beaches of Florida or the Bahamas; the rock formed is thus very different in property from the "rag" or "hardstone".

Through the intervening centuries Oxford building stone use has gone through a number of boom and bust cycles, as the qualities

opposite/ The original Earth Sciences building, adjacent to the Museum of Natural History.

left/ Ragstone blocks on St Michael's church tower.
right/ Modern oöids from the Bahamas.

of a stone variety became recognised, forgotten, over-exploited, and finally abandoned. Much of the stone for University and college buildings came from quarries in Headington, but enthusiastically inappropriate use in the seventeenth century eventually gave rise to the need for very extensive and costly renovations, and the need to find new sources.

Bath stone or Clipsham?

In the nineteenth century the predominant freestone used in Oxford was from Bath, a use brought about by the possibility of easy transport via the flourishing canal system. Bath stone, unlike any other of the stones used in Oxford, comes from underground mines. An extensive system of mines still exists in the steep hillsides around Bath, although those mines still operating occupy just a fraction of this network, the remainder having been taken over for military use during the Second World War and still used for defence purposes now. Bath stone at its best is a very good building stone as is well shown by one of the finest buildings in Oxford, the University Museum of Natural History on Parks Road which was largely built of a variety called Box Ground. Bath is an easily cut stone that has a natural advantage of allowing fine joints to be made, so fine in fact that the blocks of stone were fitted together with iron cramps rather than mortar. The attributes of Bath stone were well suited to the smooth fronted formal style of Georgian and early Victorian architecture. In a fundamental way though, Bath stone is an inside-out material. In most sedimentary building stones the individual carbonate

below/ Oxford University's Museum of Natural History.

opposite/ The Examination Schools on Oxford's High Street—an early Clipsham stone building.

grains within the rock are strong, and these are then bound together by a further strong carbonate cement; Bath stone comprises grains and cement, just like other limestones used for building, but the individual oöid grains are very weak, and unusually it is the cement that gives the rock its strength. Hence for Bath stone the nature of the cement is very important, and modern Bath stone is no longer produced from the same time-tested strata as it was in the heyday of the nineteenth century.

Towards the end of the nineteenth century and into the twentieth a new stone was introduced to Oxford–Clipsham. Clipsham stone is a variety of Jurassic limestone from Lincolnshire and Rutland. The stone was introduced to Oxford by the architect Thomas Jackson, and an early building using Clipsham was the Examination Schools on the High Street. It was soon appreciated that Clipsham was a very hard wearing stone, and new building and restorations where freestone was called for invariably used Clipsham right up to the end of the first half of the twentieth century. Clipsham reached its apotheosis not in Oxford, but instead in London, in the early twentieth century, where it was used to completely replace the original Anston stone from South Yorkshire as an external facing for the Houses of Parliament.

In the spate of recent building in Oxford, Bath stone has again tended to find most favour. The reason for this is not entirely clear, and may partly be down to good marketing, but when it came to making a decision on which stone to use for the principal west and south facades of the laboratory wing of the new Earth Sciences building, the selection of either Clipsham or Bath became a key issue, resolved eventually on the basis of history, aesthetics and technical quality. History, because the old Earth Sciences building was constructed of Clipsham; aesthetics because the facade would match those buildings on South Parks Road already built from Clipsham when viewed as a succession of imposing and varied structures from west to east; and on technical quality because of the established precedents for hard wear. An additional advantage of Clipsham is the lively texture, "cross bedding" to a geologist, which is an integral part of the blocks used in the final selection. Bath for all its creamy simplicity just cannot match up in terms of geological interest.

Stone for the narrative wall

The narrative wall provided a new opportunity to explore the further use of Jurassic limestone in an Oxford context. In particular, the technology of bonding stone to pre-cast concrete panels meant that new choices were available. Purbeck stone, essentially a 'hardstone', has been quarried in South Dorset since ancient times. As a building stone its thin beds have had a variety of uses, such as for paving, as in the Canterbury Quadrangle of St John's College, Oxford, and for church monuments such as at Oddington, on nearby Otmoor. In the new Earth Sciences building it was possible to use Purbeck in a new fashion, and the stark black Purbeck variety known as 'Feather' has been employed in simple graphic strokes on

Stone stories: from quarry to thin section: Clipsham stone (top row), Purbeck 'Feather' (middle row), Jura limestone (bottom row).

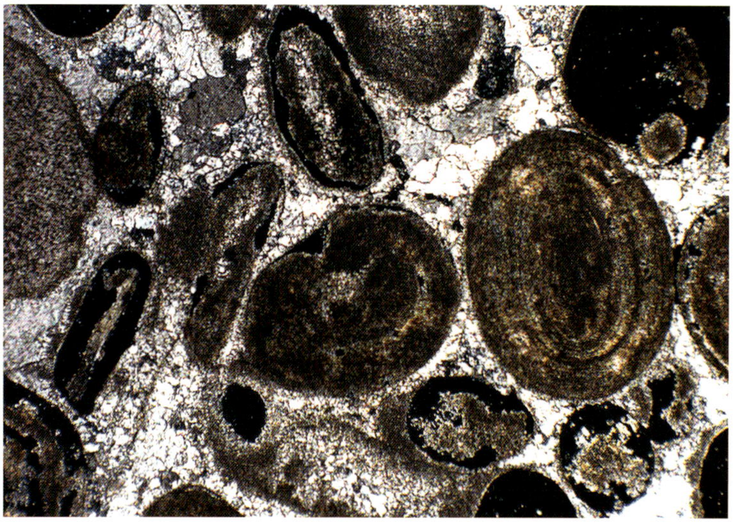

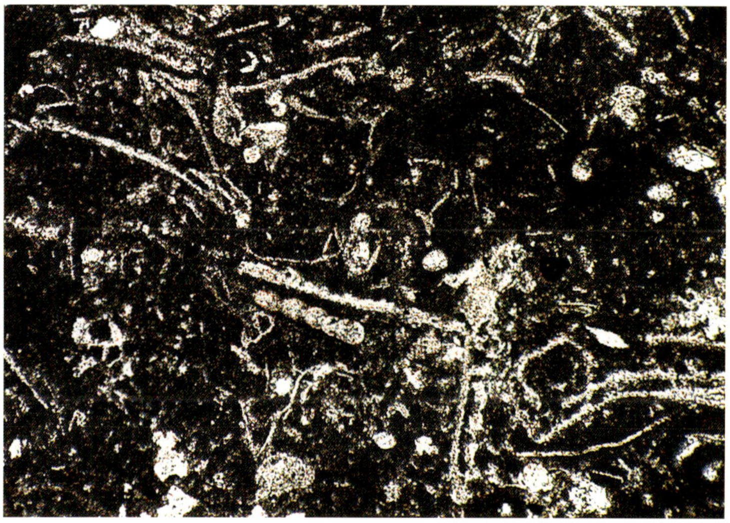

the narrative wall. The rock itself comprises concentrations of oyster shells with a very tight carbonate cement. Purbeck stone is nearly always shelly and was laid down in extensive lagoons and lakes on the edge of the Jurassic seaway. The lagoons in which the original sediment was laid down were evidently shallow, shallow enough for there to be dinosaur footprints, though not yet found in any of the slabs used in the new building.

The major part of the narrative wall is Jura stone, from Bavaria. The Jurassic strata from this part of Germany are most famous for providing the bird-like reptile fossil archaeopteryx, from the Solnhofen limestone, but quarries in the underlying beds yield a good hard limestone suitable for building. Jura stone has been much used recently for a variety of large buildings in London; generally, though, the surfaces have been tooled in a variety of styles, some of which evidently serve to make it look more like Portland, the building stone of choice in London if you wish to make an emphatic architectural statement. For the new Earth Sciences building we wanted to take full advantage of the great range of different fossils present in Jura limestone, and for this reason selected blocks that were particularly rich in fossils, and finished with a smooth surface. These fossils include those typical of the Jurassic, ammonites and belemnites, and also some less familiar in an English context including sponges. Two varieties have been used, a pale cream variety with abundant fossils, and a pale greenish brown variety in which the traces of burrowing animals are clearly visible.

Basement, paving, traps

Care and attention has also been given to the paving stones inside and out. "Basement" is a term that means something distinct to a geologist; it is the hard, usually coarsely crystalline rock, formed relatively deep down in the crust, on top of which sedimentary rocks are later laid down. Basement is commonly granite, a rock formed by slow crystallisation from a melt. For the paving outside the new Earth Sciences department, forming a metaphoric basement for the whole building, a clean white granite from northern Portugal has been used: Amarelo Real Granite, which in nature forms the basement to the younger sedimentary strata. Inside, the paving has been continued in Pietra Serena Sandstone, a remarkably homogeneous medium grey sedimentary rock from Tuscany, made up from very well-sorted, fine-grained quartz sand and mica. At the other end of the grey scale is the geologically black and very young Italian basalt that forms the nosings and landings for the main staircase. Trachyte is a fairly common finely crystalline rock produced by the solidification of sheets of molten lava extruded onto the Earth's surface. Supplied as "basaltite", our trachyte is rather unusual chemically, being rich in potassium. With the trachyte there is another subtle link between the geology and the architecture. In Scandinavian languages the stem word for stairs is "trap", and stepped landscapes have formed by the differential erosion of vast sheets of solidified lava flows, such as is the case in Iceland. Hence in Geology large accumulations of successive lava flows are called "Traps". What more appropriate stone could have been found for the stairs in an Earth Sciences building?

I am particularly grateful to Tim Palmer, of the University of Wales in Aberystwyth, and Philip Powell, of the Oxford University Museum of Natural History, whose knowledge of traditional building stones is far greater than mine, and whose enthusiasm for the subject has stimulated my own interest. I would also like to thank Owen Green and Jeremy Hyde for their help in producing the thin sections that illustrate this piece.

Stone stories: from building to thin section: granite (opposite column), trachyte (left column), Pietra Serena (right column).

Section 3/ Developing a design

Chris Wilkinson

The Oxford condition

For an architect, working in Oxford is both inspiring and also a little daunting because of the wealth of great buildings, the power of the planners and the academic nature of the inhabitants. In Oxford, like many other historic cities, architecture is the subject of much debate. There is a keen appetite for the preservation of the historic environment and concern about how new buildings should be introduced alongside the older ones. Any new design has to pass considerable public scrutiny which can lead architects to take the safe option, and whilst this is appropriate in some instances, there are sites where a clear statement of modern architecture is the right solution. Good architecture has the ability to reflect aspects of our lifestyle, culture and priorities so it is necessary to be able to read the period of construction with clarity.

In general, this has happened at Oxford, with its rich mix of Classical, Gothic and Modern architecture. Strong interventions, such as William Butterfield's Victorian Keble College, sit comfortably amongst the more traditional stone spires and perpendicular carvings. Sir Giles Gilbert Scott's New Bodleian Library of the 1930s is perhaps less popular but is, nevertheless, accepted in its distinguished historic surroundings. Arne Jacobsen's much admired St Catherine's College of 1960 established a new modernist style for Oxford buildings, a trend continued by Powell and Moya at Wolfson College in the 1960s, and followed by Arup's Sir Thomas White Quad at St John's in 1975. James Stirling's Florey Building of 1966 took a bold, uncompromising position which was highly controversial and has led to a more contextual approach for subsequent architectural commissions in the city. This has taken the form of well mannered Modernism such as the work at St John's College by MacCormac Jamieson Prichard on the one hand and Demetri Porphyrios's classical theatre at Magdalen College which reverts to pastiche.

On our commission for the Earth Sciences building, therefore, one of our main concerns was how we could develop an appropriate architectural expression for the building, squeezing the demanding requirements of the department's brief onto a constrained site whilst still respecting the historic urban context. Our approach was to design a functional building which is very site specific, uses modern technology in its construction methodology and reflects some of the idiosyncratic character of the Earth Sciences department in its architecture. Our priorities were to use massing, scale, proportion and quality of materials as the key to fitting into the local context. We have confidence in Modernism and feel comfortable mixing the new with the old in the belief that time is an important marker.

Our discussions with the planners were positive from the start and the design was agreed through a process of consultation, culminating in a successful exhibition in the Museum of Natural History. Although the architecture of the building is unashamedly contemporary, we found that people generally responded well to the concept and enjoyed the narrative.

The first conceptual ideas were formed at competition stage and many of them remain today, in the built scheme. They are anchored in a clear organisational diagram in which the

clockwise from top right/ James Stirling's Florey Building at Queen's College; Demetri Porphyrios at Magdalen College; St John's College; Wolfson College; Arne Jacobsen at St Catherine's College; The New Bodleian Library; The chapel at Keble College.

New and old: the Earth Sciences building sits between the Dyson Perrins Laboratory and the Department of Inorganic Chemistry.

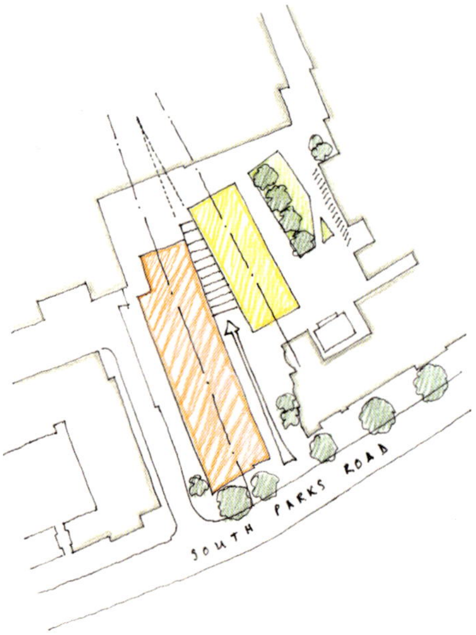

top/ The organisational diagram separates the laboratories from academic and administrative offices.
bottom/ The Clipsham-clad laboratory wing.
opposite/ View from South Parks Road, showing laboratory and office wings separated by the central atrium.

laboratories are housed in a stone-clad rectangular wing set orthogonal to South Parks Road with the academic offices and administrative activities accommodated in a smaller, glazed wing, set back from the road on axis with the adjacent Dyson Perrins building. These two elements are joined by a service core and an atrium housing the main entrance and reception. This geometry was found to be the most effective way of incorporating the defined areas of the brief on such a tightly constrained urban site, whilst still allowing access through Le Gros Clark Place to the University's Science Area beyond.

The building also responds to its context on the western boundary of the Science Area adjacent to the historic city core. The laboratory wing, for instance, has a strong presence on South Parks Road, but also presents a long facade onto Robinson Close behind the historic Pitt Rivers Museum and the Inorganic Chemistry Building. It has therefore been designed with a rather formal appearance to relate to these nearby buildings and is clad in Clipsham stone inset with large modular window openings. Clipsham is a Jurassic limestone from Lincolnshire that has been used on many of the historic buildings of Oxford and was the preferred choice of our well informed client. The scale of the windows has been designed to provide good daylighting to the laboratories, while also making reference to the neighbouring Dyson Perrins Laboratory. The layout of windows generally follows the regular arrangement of laboratories within, but

with some modules missing and some standing proud to break the rhythm and to make the composition interesting. Above the flat parapet, which is set at the maximum permitted Carfax height, stand the flues which are designed as an integral part of the architecture, making a distinctive vertical statement on the skyline.

The Carfax height, which is defined by the Carfax monument on Oxford's High Street, generally sets the maximum permissible height for new development in the city at 79.3 metres above sea level; but on the Earth Sciences building, consent was given for the flues, plant room and top floor Senior Common Room to exceed this height, provided they were set back from the main facades.

One of the consequences arising from this building layout is that the main entrance lobby is set back quite a long way from the main road. In many ways, this is entirely consistent with the 'Oxford condition' in which many of the college entrances are approached through courtyards. In this instance, the entrance reception is well placed centrally between the laboratory wing and the office wing. It serves to enliven Le Gros Clark Place but leaves an expanse of the east elevation of the laboratories exposed. This has become the building's "narrative wall", an architectural device that helps to draw visitors from South Parks Road to the entrance.

The idea for the narrative wall was proposed at competition stage to give the Department of Earth Sciences the opportunity to express their subject on the outside of the building. It was embraced enthusiastically by Faculty members and has proved to be a lively talking point for all those who interact with the building, as can be seen elsewhere in this book.

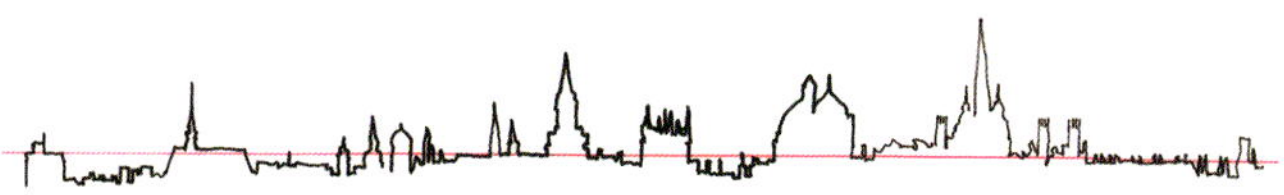

opposite/ View back to South Parks Road along the narrative wall.
top/ The bulk of Oxford's buildings lie below Carfax height of 79.3 metres above sea level, although this line is punctured by the strong verticals of the City's famous spires.
bottom/ View of the Oxford skyline from the Senior Common Room.

In contrast to the stonework cladding to the laboratory wing, glass was selected for the office wing. Here, the light, crystalline appearance of the building enhances the rather intimate space of Le Gros Clark Place. Although the material treatment is very different, the language of this wing is the same as the Robinson Close elevation, with large squarish windows set into a glass panel glazing system. The upper part of the wing at the south end is cantilevered over the courtyard, two floors up, to improve access and to keep the space light and open. This is important for the student lecture theatres at ground level which face onto the courtyard.

Our response to this courtyard was informed by the masterplan for the Science Area, prepared by Sir Terry Farrell, which placed a high priority on connectivity and the quality of public open space. The plan of the new Earth Sciences building makes maximum use of the room available whilst retaining the use of Le Gros Clark Place as a space in which to linger for recreation or to pass through into the hinterland of the Science Area.

Special care has been taken in the design of the landscaping to fulfil these activities and to create an appropriate identity to enhance the new Earth Sciences building and the Grade II listed Dyson Perrins Laboratories. The hard landscaping, for instance, uses natural stone pavings laid with an angular pattern that refers to the geometry of the narrative wall. A large, invasive cycle storage hut has been demolished in order to open up space for a formal lawn and rationalised cycle provision along the eastern boundary and within the basement of the new building. Soft landscaping has been planted to areas of the perimeter with trees included where possible, and a small, almost rectangular formal lawn has been grown within the centre of the main courtyard space on a geometry related to the building's office wing. In tune with the narrative for the Earth Sciences department, a line of Ginkgo trees–the oldest tree species, known as the "living fossil" and believed to date back some 270 million years–has been planted on the lawn.

Whilst the soft landscaping will take time to mature, it is already possible to appreciate the intimate scale of this new public space and it is pleasing to see how the new building relates to the existing Dyson Perrins building, which now seems to have gained a more positive presence. The Science Area as a whole has benefited from this new gateway space, which draws people in from South Parks Road and provides an appropriate setting for the contemporary architecture of the new Earth Sciences building within its constrained urban context.

left/ View from Le Gros Clark Place showing cantilever of office wing.
right/ Landscaped courtyard.
opposite/ Contrast of glass and stone in the atrium.

Stafford Critchlow

No stone unturned

"Tectonics", the title of this book, is a term that appears in the lexicon of both Geology and Architecture. For geologists it is associated with the formation of mountain belts; for architects it is "a means to reveal the essence of building",[1] the concept that ties all the elements of a project together,[2] or a way of describing the transition between component parts from which the beauty of architecture emerges.[3] Both subjects are concerned with scale and jumping between the macro and the micro: for geologists, going from cliff face to thin section of stone under the microscope; for architects, from cityscape to detailed junction between materials. Perhaps it is not so surprising that Geology was a taught curriculum subject on the first Architecture course at University College London in 1895.[4]

In Oxford, the famous Natural History Museum marries innovative glass and iron architecture with exhaustive representation of geological specimens in the manner of Victorian educational philanthropy. The Earth Sciences department, previously known as Geology, has moved from a location adjacent to this building and can lay claim to some of its collection.

In designing a new building for Earth Sciences in a city synonymous with the ochre stone of the Bodleian library and the Radcliffe Camera, the design team were presented with an exceptional opportunity to express the synergy between client, context and materials. A number of Wilkinson Eyre's recent projects have sought to find an architectural character drawn from the identity of the building's users. The chance to allude to geological layering and represent the preoccupations of the

department in the architecture was a key aspiration when we submitted the competition design in September 2005.

In architectural terms (as well as establishing a clear diagram for the building with separate laboratory and office wing with an atrium between) we wanted to set out our stall with a narrative about the building occupants. The so-called "narrative wall"–the long elevation that leads one from South Parks Road to the entrance and into the atrium–is an organisational device that separates the two wings of the building while offering a subtle reminder that you are entering the world of Geology. From a distance geologists may consider it much as they might a cliff face. Closer up, it can also function as an educational tool. Fossils–belemnites and ammonites that occur in the Jurassic limestones selected for the building–are present in the lower levels of the facade, and can be examined as specimens.

I was fortunate to be invited on the undergraduate field trip to the volcanic islands of Santorini shortly after we were appointed to work on this project, to get a feel for the subject, the people involved in studying, teaching and researching it, and to start to draw out the unwritten brief for their new building. People were patient in explaining technical terms and afterwards I sent a portfolio of my photographs to Philip England. It amused him, I think, that these photos were quite unlike ones that a geologist would take, and this started a conversation which continued throughout the project. I was struck by how the dramatic story of the volcano emerged over a number of days through a detailed reading of the cliffs around the caldera;

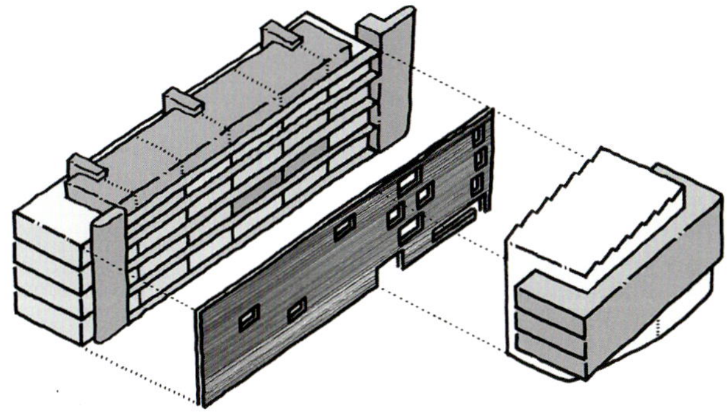

opposite left/ The pillars surrounding the main court in Oxford's Museum of Natural History are each constructed from a different British stone.
opposite right/ The museum's famous glass and iron architecture.

top/ The basic building elements: laboratory wing, narrative wall and office wing.
bottom left/ Reading the landscape on the Santorini field trip.
bottom right/ Sketch showing how the building's plan facilitates interaction between lab and office space.

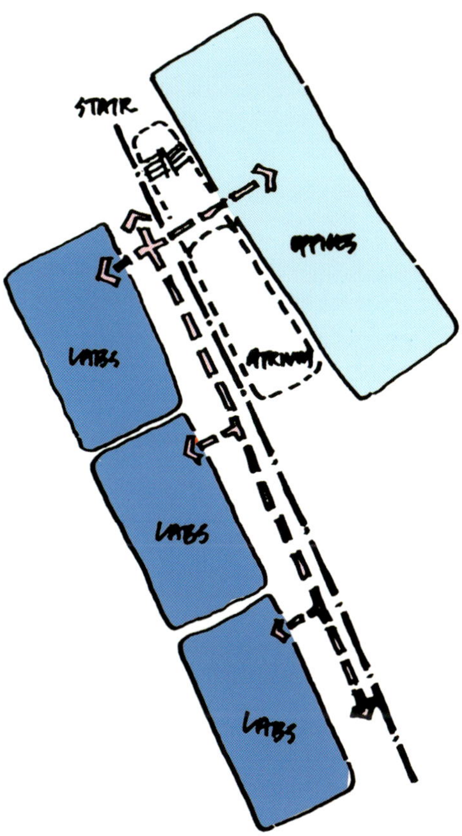

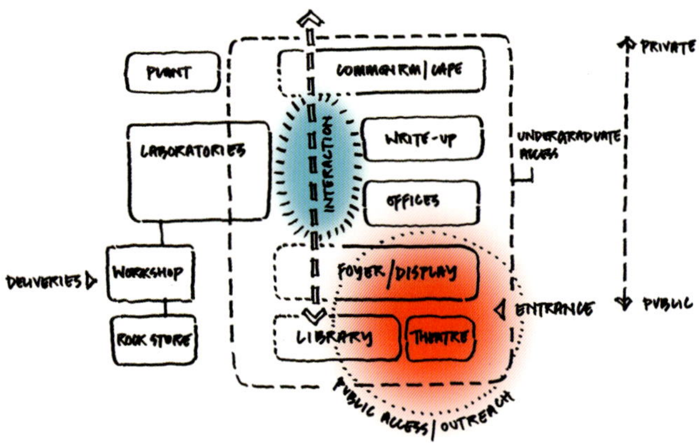

top and bottom/ Sectional concept sketches showing diagram, zoning and relationship of floors to Carfax height.

opposite/ Physical models were used to develop the design.

how visually contrasting features demonstrate the magnitude and timing of what happened; the importance of scale on perception—that what appears at a distance as one thing is quite another close up. And intertwined in the geological story of successive eruptions is also a human one: tantalising speculations over Santorini's role in the legend of Atlantis and the end of Minoan civilisation.

And so the design of the narrative wall was developed as linear horizontal bands that rise and fall like strata on a cliff face. Two colours of Jura limestone—beige and gold—are articulated with different relief, the gold being 50 millimetres thicker; and spliced through the whole are tapering ticks of dark Purbeck feather (only available in small quantities from certain quarries in Dorset—but the department felt strongly that the stone of the Jurassic Coast, frequently visited on field trips, should play its part here).

Many people have explained to me that Earth Sciences is science for hill walkers. The characters who choose it as a discipline tend to be outdoorsy. They are interested in a relationship with the natural world and their new building, while offering state-of-the-art research facilities, should not look or feel 'clinical'. Accordingly, while the planning of the building is rational and logical, the narrative wall running through the atrium is an important foil to the controlled scientific environment: it contradicts the hegemony of science research buildings by being particular, not generic. People catch a glimpse of it from their desk or as they move around the building, a reminder

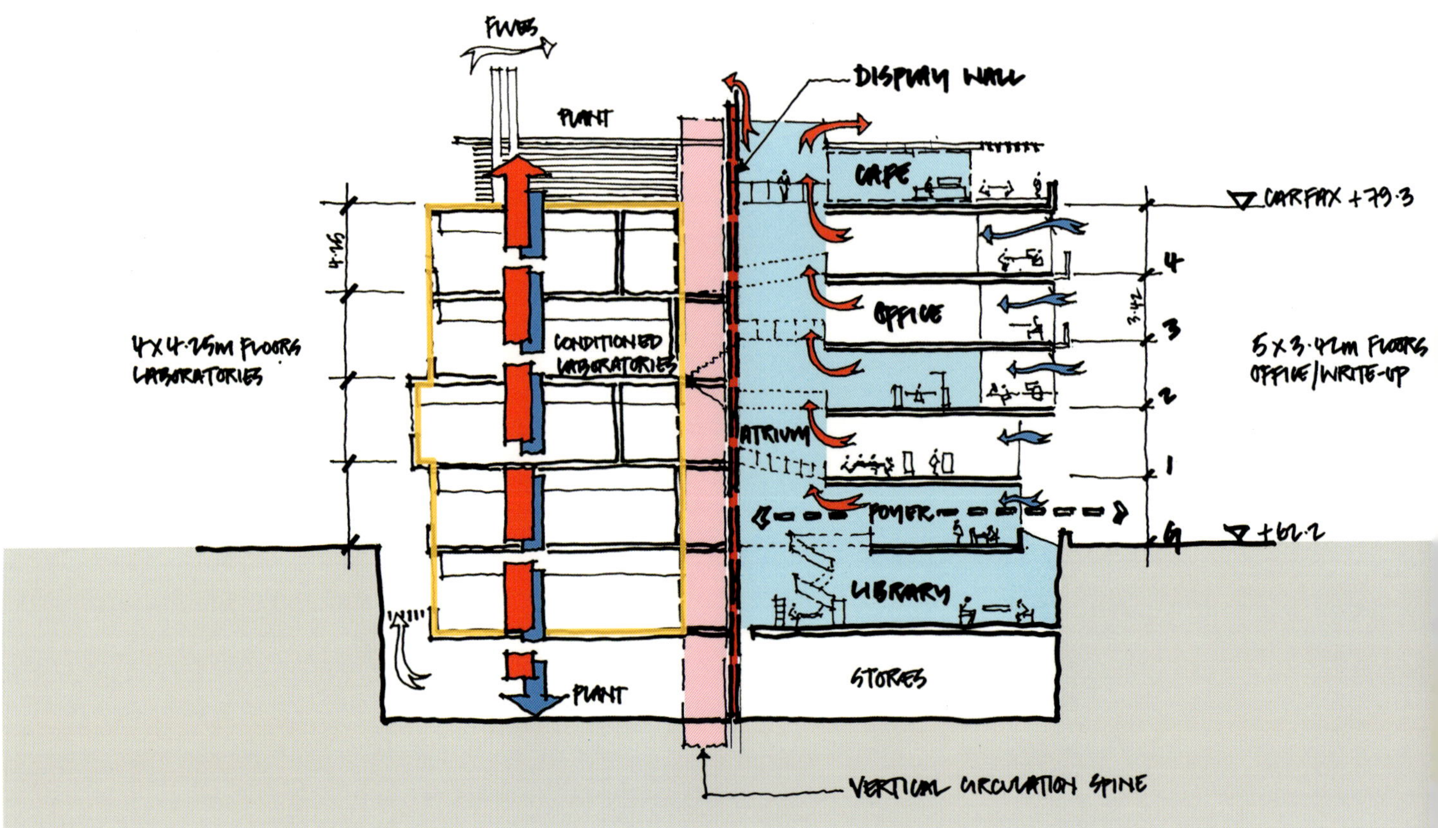

that despite high-tech innovations, one foot is still planted firmly in the natural world.

The field trip also demonstrated the close interaction between undergraduates and the academic staff. Communal coffee breaks, a feature of life in the old building, are honoured in the new one with the senior common room on the roof providing a meeting place for academics. The faculty's interdisciplinary nature is one of its most striking characteristics and it is received wisdom that many advances in scientific research–eureka moments–are triggered by informal discussion amongst academics comparing specialised strands of research.

The plan arrangement of the building was largely dictated by the site constraints. Planning restrictions limiting new buildings to Carfax height meant that by separating laboratories and office space into two separate wings we could optimise the accommodation provided, with five floors of office and four floors of laboratory, within the pre-ordained overall building height. This set up a level change between the two wings that is resolved around the lift core but also provides spatially interesting half-level changes as one moves about the building.

Set back above the parapets at Carfax height are rooftop pavilions: a large plant room above the laboratory wing–five metres high but positioned behind a stone parapet wall to conceal its bulk and articulated as three distinct louvred elements with clerestory glazing and a floating zinc roof above–and on top of the office wing a glass-walled senior common room with floating roof and brise soleil extending across an outdoor decked terrace and green roof.

Anchoring the laboratory wing roofscape are four chimneys that march along Robinson Close perpendicular to South Parks Road. These contain the fume extract flues and are designed at an angle to balance the mansard shape on this side of the roof, the lower portion widening to accommodate the intersection with the plant room. The filigree louvres to the upper part are wrapped on three sides by solid render to slim the proportion and enhance verticality. The planners were supportive of these vertical elements puncturing the uniformity of the Oxford skyline. For the planning application process, as well as considering street level views along South Parks Road, we also prepared computer verified views to study the new building's visual impact from key distant views such as the one illustrated overleaf from St Mary's Church. The chimneys were made of steel in two halves off-site and craned up onto the building. They are clad in self-coloured acrylic render, a couple of shades darker than the Clipsham stone on the west elevation.

The diagram of the building as two distinct wings also supported a couple of other strategic objectives: that the scientists should move about the building during the day between academic offices and laboratories, thereby encouraging interaction (the landings facing into the atrium are generously sized to encourage this); and that, by being set back from the noise of South Parks Road and facing onto the traffic-free Le Gros Clark Place to the east, the office wing could be naturally ventilated

top left/ Planning view from St Mary's Church. The building sits mid horizon between the Radcliffe Camera and the Hans Krebs tower, now scheduled for demolition.
top right/ Planning view along South Parks Road.
bottom/ Outdoor terrace to the senior common room.

opposite/ Rooftop flues to the laboratory wing.

by opening windows with user-operated external blinds to screen out early morning sun, thereby reducing overall energy use.

The plan arrangement on the ground floor follows the logic of separate office and laboratory wings with undergraduate and conferencing facilities overlaid. The entrance to the building is straight into the small atrium. Here is reception with administrative offices behind, seminar rooms to the right and the undergraduate common area and teaching labs to the left. The narrative wall tapers off into the ground as one approaches the entrance and therefore the full-height glazing onto the undergraduate common areas and displays on the back wall can act as something of a shop window for Earth Sciences. To the south, and fronting also onto South Parks Road, is a small departmental library containing key resources—map tanks and journals—as well as providing a quiet study space.

The upper floors of the office wing are organised as cellular academic offices of between 12 and 16 square metres (large enough to accommodate a small meeting table for tutorials as well as a desk and bookshelves) ranged along the perimeter of the building with good natural daylight. The occupant of each office controls their own environment with an openable window and electrically-operated external blind for shade. Internally, larger four-person offices for post-doctoral researchers face into the atrium with partial height glass screens up to two metres high to enable natural ventilation. At

top left/ Basement workshop.
bottom left/ Academic office.
right/ Metal-free laboratory.

opposite/ Seminar room.

either end of the wing are group working areas for six used by fourth-year undergraduates who are beginning to undertake academic research.

The laboratories on Levels One, Two and Three all have good daylight and are organised to suit the specific requirements of each research group either side of a grid of riser shafts and a central corridor. On Level One the corridor is offset to provide larger 70 square metre laboratories for culturing, environmental microbiology and experimental petrology to the west side. On the floors above, the corridor is central with more equally sized labs ranging between 20 and 60 square metres to each side. The suite of ten metal-free Geochemistry laboratories on Level Three accessed via gowning chambers and a clean corridor presented probably the most technically challenging spaces to build–all ironmongery is plastic, the windows are lined with plastic-framed secondary glazing and the walls covered in welded vinyl. On Level Two the geochemists have the largest suite of university mass spectrometers in the world (13 in total)–positioned away from the lift core to mitigate against possible vibrations. In the large corner lab one of the windows was designed to be removable so that the largest of the mass spectrometers could be lifted in by crane post-occupation.

In the basement are general workshops served by a hoist in the lightwell off Robinson Close and specific workshops for rock crushing and creating thin sections of rock for study under a microscope. Also on this level are laboratories that do not require daylight: the palaeomagnetism laboratory and the field-free room (one of only three in the UK) where all magnetic fields have been eliminated.

The composition of the elevations was developed to achieve our aesthetic aspirations with the constructional techniques available to us. The laboratory wing is clad in large panels of precast concrete faced with 40 millimetre thick Clipsham stone blocks arranged in 400 millimetre high courses with randomised vertical joints of 300, 450 and 600 millimetres. A larger scale jigsaw of the precast panels, which have mortar coloured silicone joints between them, was designed along the facade that appears as an overlaid pattern relating to the interlocking horizontal and vertical subdivision of the large three by three metre square windows. These windows themselves have a shift in spacing between floors along the facade to create variety as they run along the long facade to Robinson Close and wrap around the corner. There are also a sprinkling of projecting oriel windows to relieve the otherwise sheer face of this wall. The office wing is clad in opaque glass panels and has square windows in a similar arrangement to the laboratory wing, though all reduced slightly in scale to 2.4 x 2.4 metres due to the smaller storey heights.

The design of the narrative wall was an exercise in distilling an apparently complex geometry into as few panel types as possible. The mould for each 7.2 metre long chamfered precast panel with stone facing was repeated, mirrored or reused by masking some of it off to minimise the number of different panel types. We chose to run horizontal stone coursing with random

vertical joints as we had done for the west elevation: visually, this seemed especially important here where top and bottom line were angled. The ribbon windows between bands of stone are shuffle-glazed with projecting angles at the top and bottom to give a shadow line and flush silicone joints between panes to emphasise horizontality. At the top of the wall against the sky a horizontal steel channel contains the jagged geometry when seen against the sky.

The narrative wall may at first glance seem random, but it has an underlying rigour of geometrical setting out that becomes apparent on closer observation. The richness of detail implying crystalline geological structures and sedimentary layers seemed to us important for this building in making it bespoke and, as the client wanted, more than just a 'clinical' science research building.

1. Karl Botticher's treatise on Tectonics quoted in *Digital Architecture and Construction*, Ali A & Brebbia C A, Southampton, 2006.
2. Kenneth Frampton, ibid.
3. Gottfried Semper, ibid.
4. See *Architecture–art or profession? Three hundred years of architectural education in Britain*, Crinson, Mark and Lubbock, Jules, Manchester, 1994.

Design process: Clipsham panels

left/ The laboratory wing is clad in precast concrete panels faced with Clipsham stone. The stone is designed to read as an interlocking 'pinwheel' pattern around a series of large squares, most of which are punched through as windows. The length of the concrete panels was limited to allow for ease of transport to the site. However, the pinwheel design had to be reflected in the shape of these panels to ensure that the joints lined up and the pattern could still be read in the final wall. These diagrams show the setting out for the concrete panels so that this could be achieved using the minimum number of panels, repeated to avoid wastage.
right/ On the completed facade, the difference between the panel and pattern joints is barely perceptible.

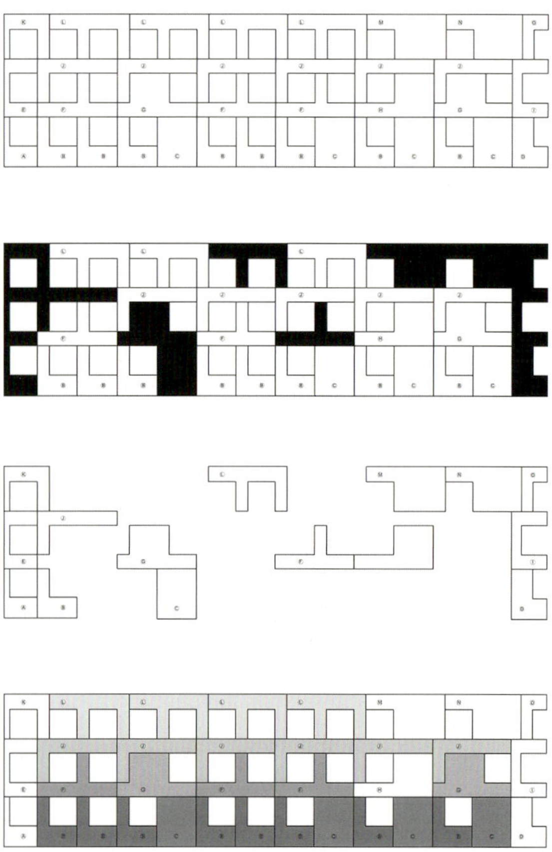

Design process: narrative wall

These early concept sketches for the narrative wall demonstrate the complex stratigraphy that can be achieved using the folded planes of the stone panels, the bands of glazing between and the texture of the stone.

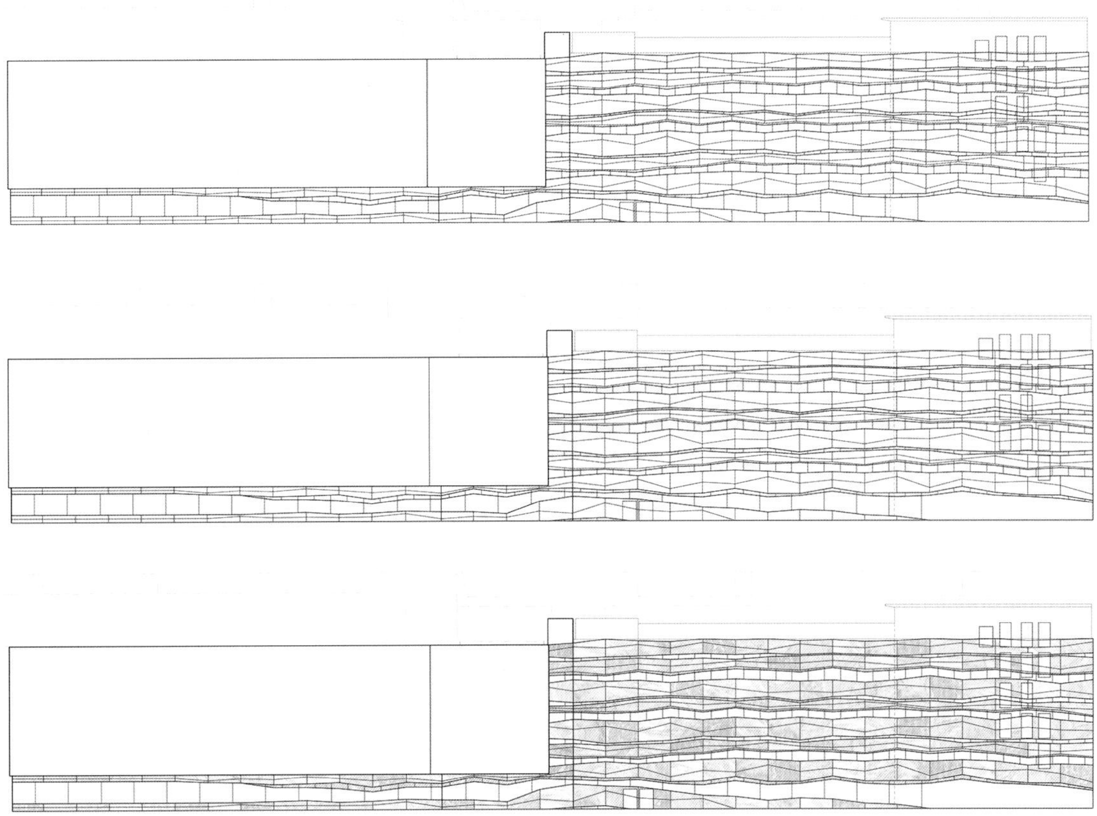

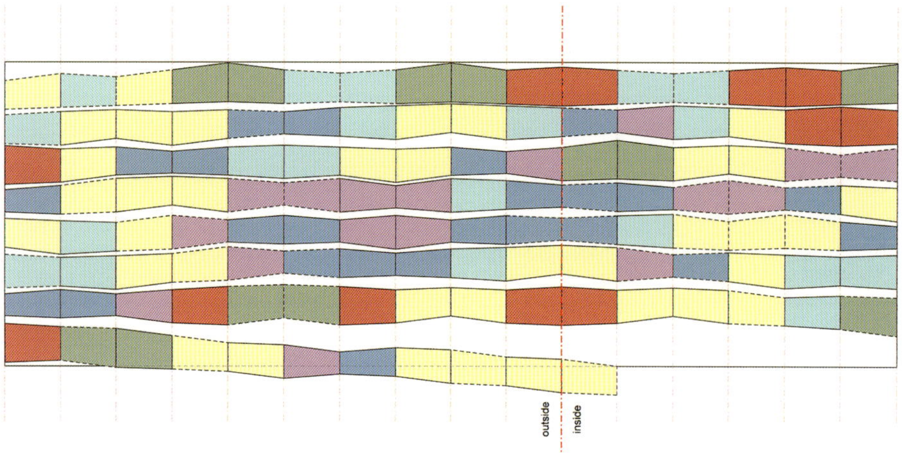

Detailed studies were carried out to examine how the best effect could be achieved using the smallest number of panels, particularly at the early stages when the wall was designed to be fully three-dimensional. This sketch shows how just six panel types could be used both in mirrored pairs and in sequence to maximise repetition and in turn ensure efficiencies in construction.

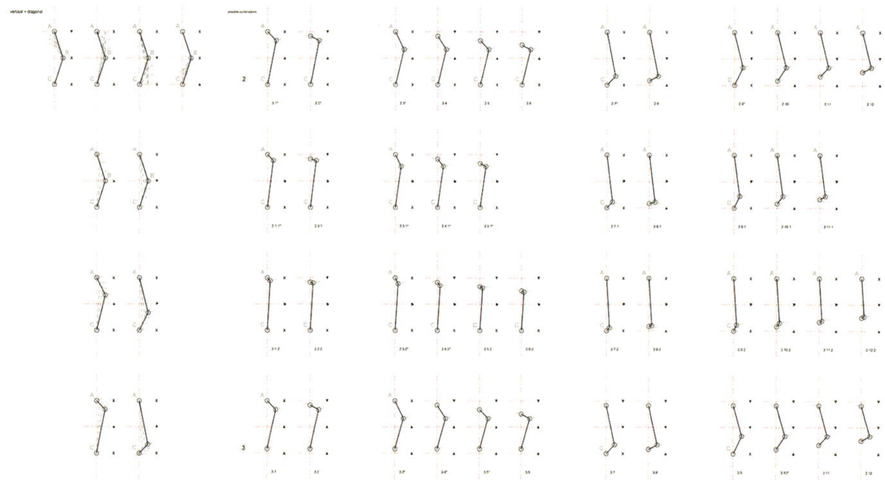

How the end section of each panel fitted to the next was a critical issue, as shown in this sketch which explores how the section influences the profile and shape of each fold.

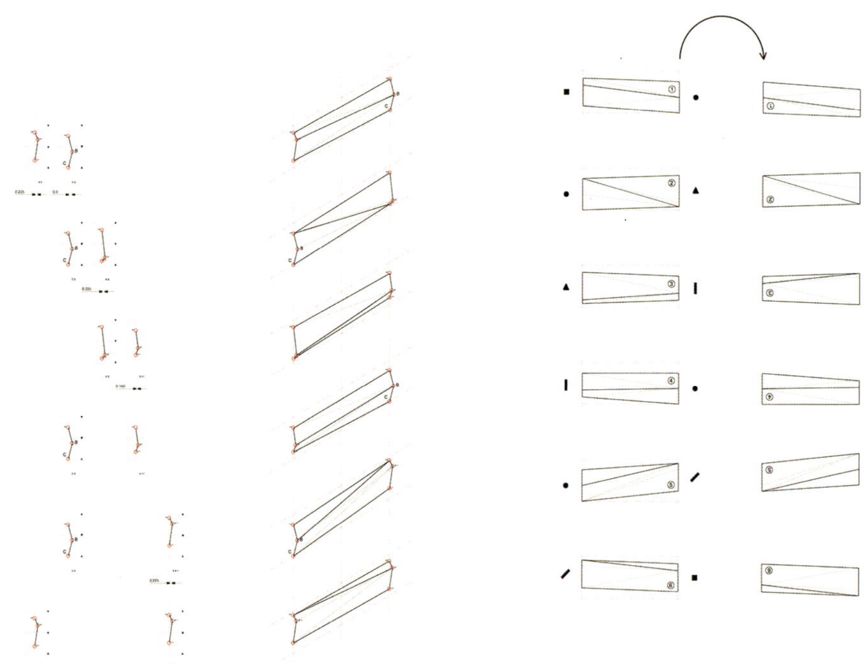

Limiting the design to six panel types ensured that the panel ends would match as well as offering a large variety of effects depending on their composition across the wall.

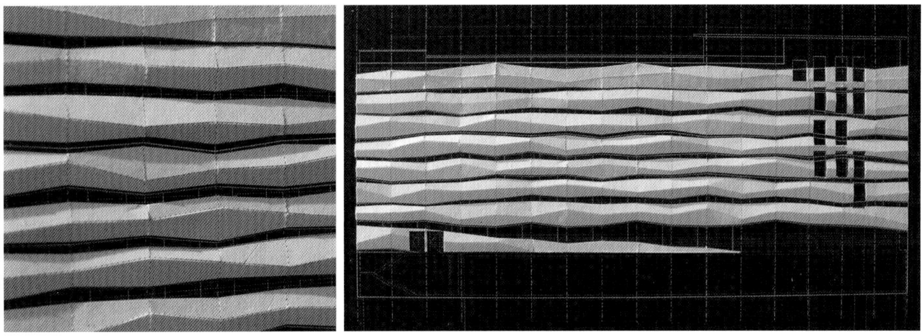

Study models demonstrated the effect of shadow and light across the folded planes of the wall.

The team developed a more cost effective two-dimensional scheme for the wall (retaining a three-dimensional strip at ground level) as the design progressed. This concept sketch demonstrates the visual effect of the panelisation if the wall is ‘unwrapped’.

These studies show the composition of the various material elements—stone types and thicknesses, glazing, fritting and so on—within the wall which, when combined, create a richly textured effect.

The shop drawings produced by Malling, the subcontractor for the narrative wall panels, and especially those for the extremely complicated 3-D panels at the base of the wall, are resonant of the colourful geological maps used within the department.

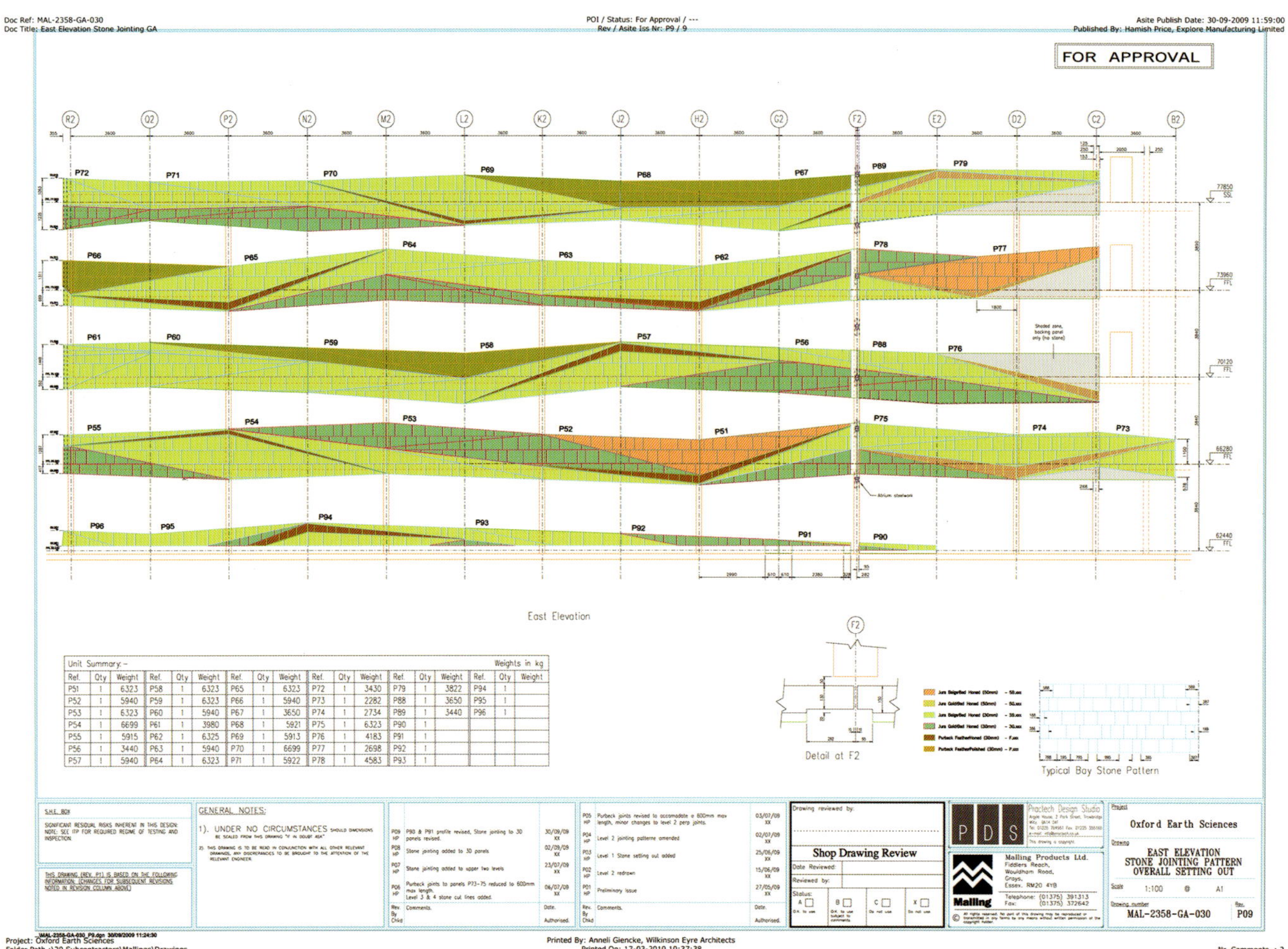

Detail of the wall as built showing the layered effect achieved through combining the variously coloured limestones.

Drawings

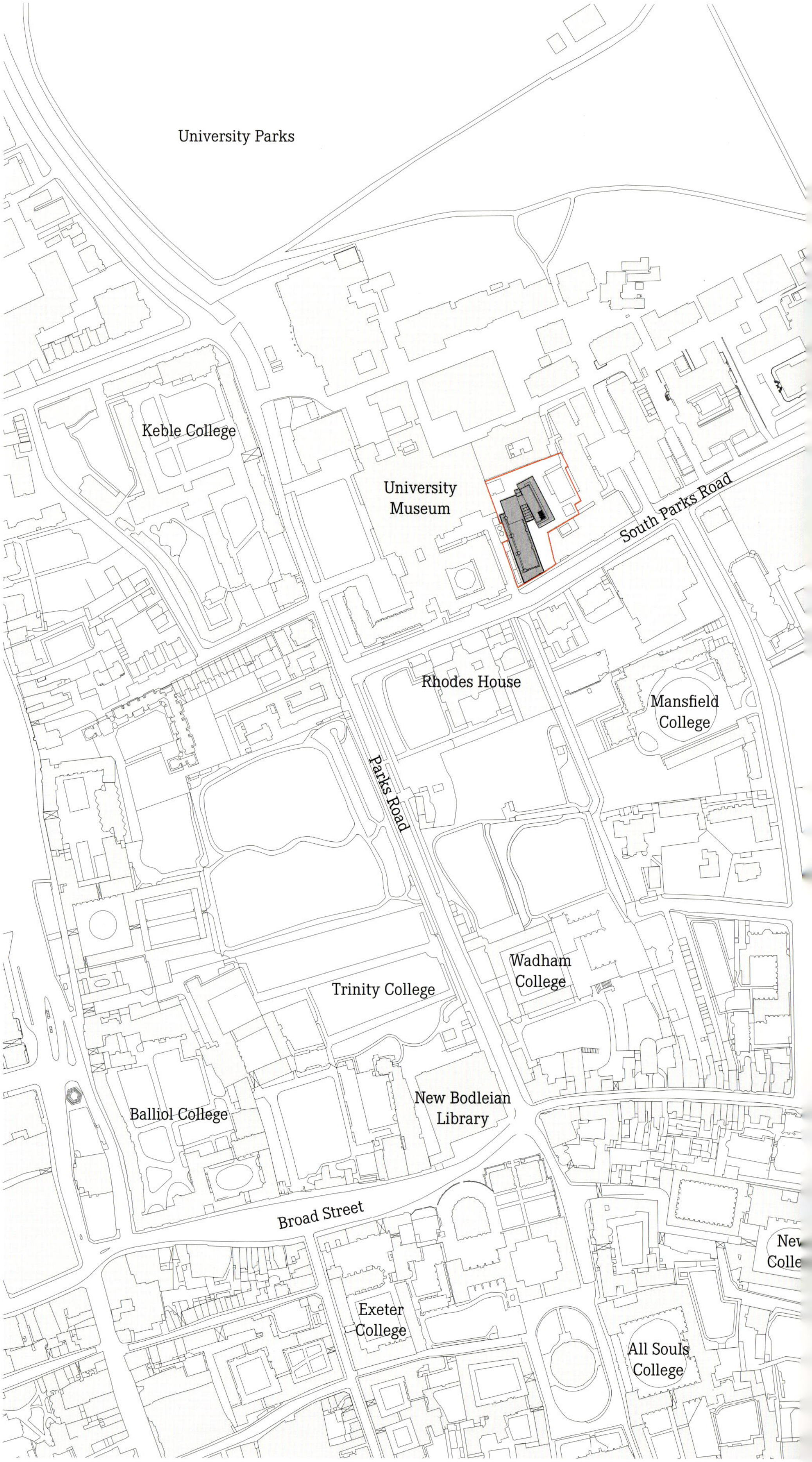

right/ Location plan.
opposite/ Site and landscape plan.

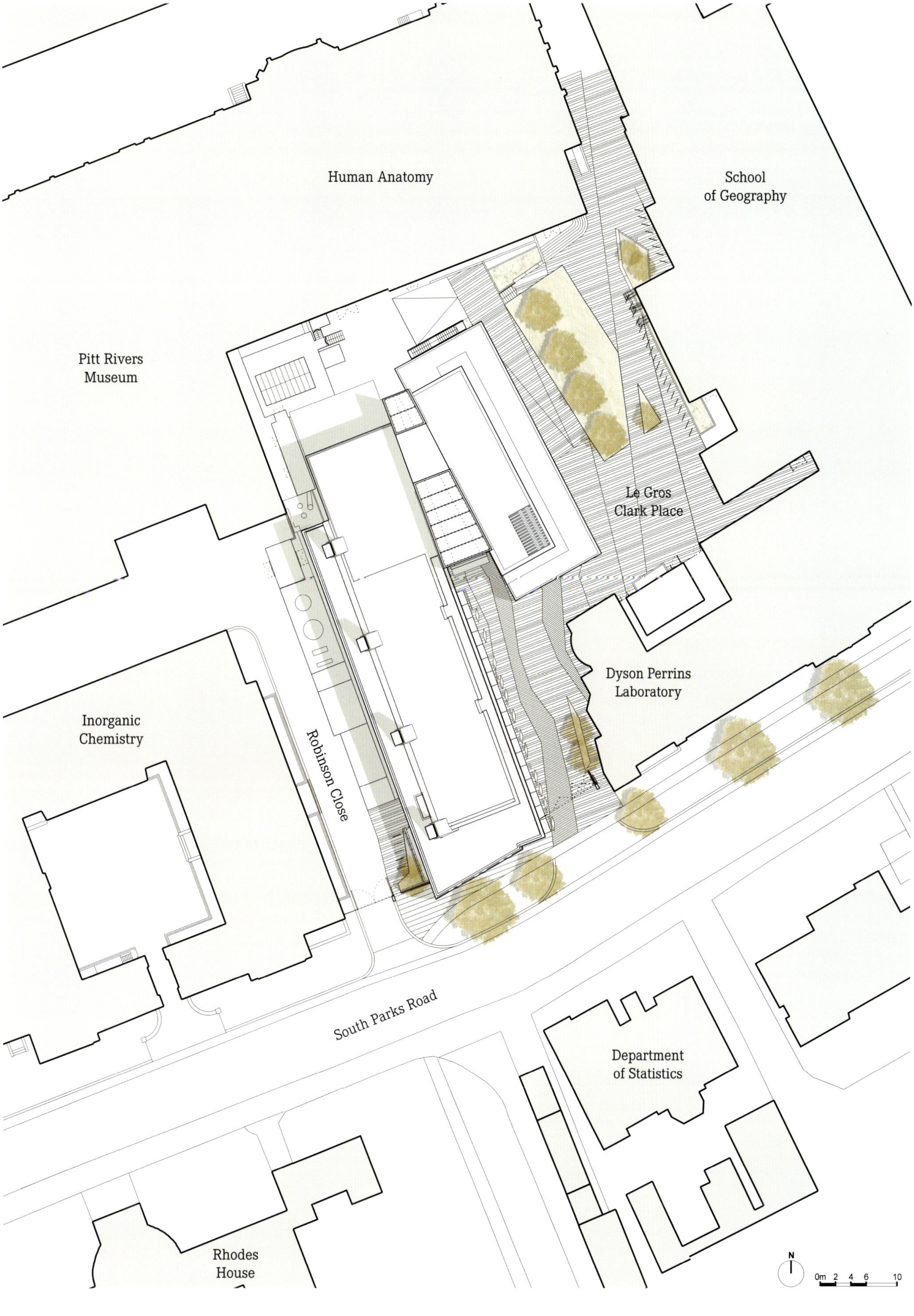
Human Anatomy
School of Geography
Pitt Rivers Museum
Le Gros Clark Place
Dyson Perrins Laboratory
Inorganic Chemistry
Robinson Close
South Parks Road
Department of Statistics
Rhodes House
N
0m 2 4 6 10

Ground floor plan

1/ reception
2/ seminar rooms
3/ administrative offices
4/ undergraduate common area
5/ teaching laboratories
6/ library

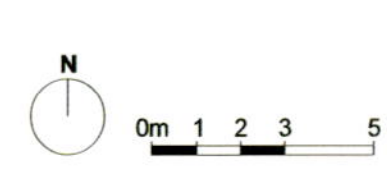

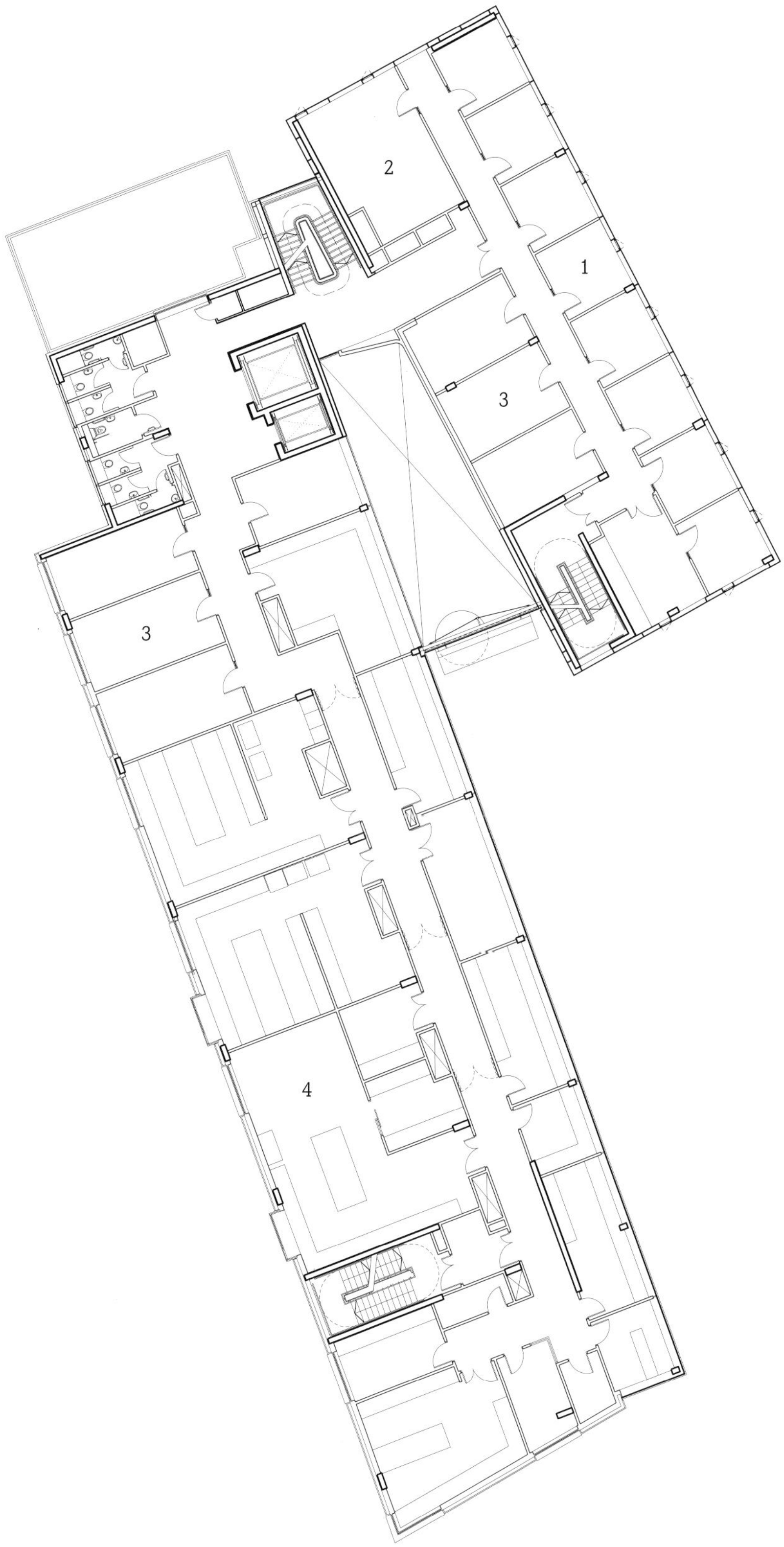

Level One plan

1/ academic offices
2/ group working area
3/ offices for post-doctoral researchers
4/ general laboratories

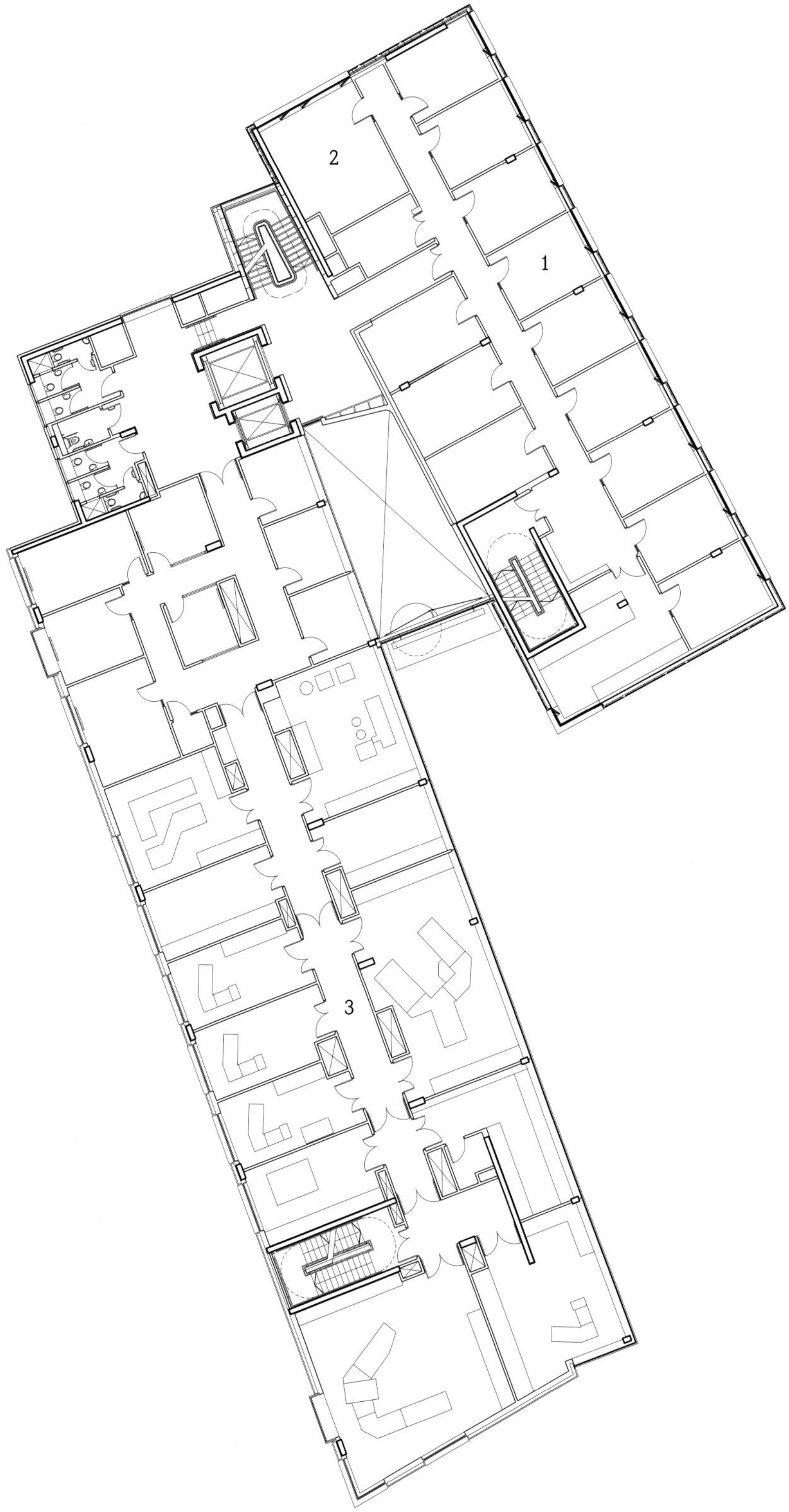

Level Two plan

1/ academic offices
2/ graduate student office
3/ mass spectrometry suite

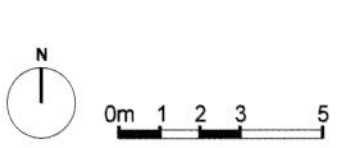

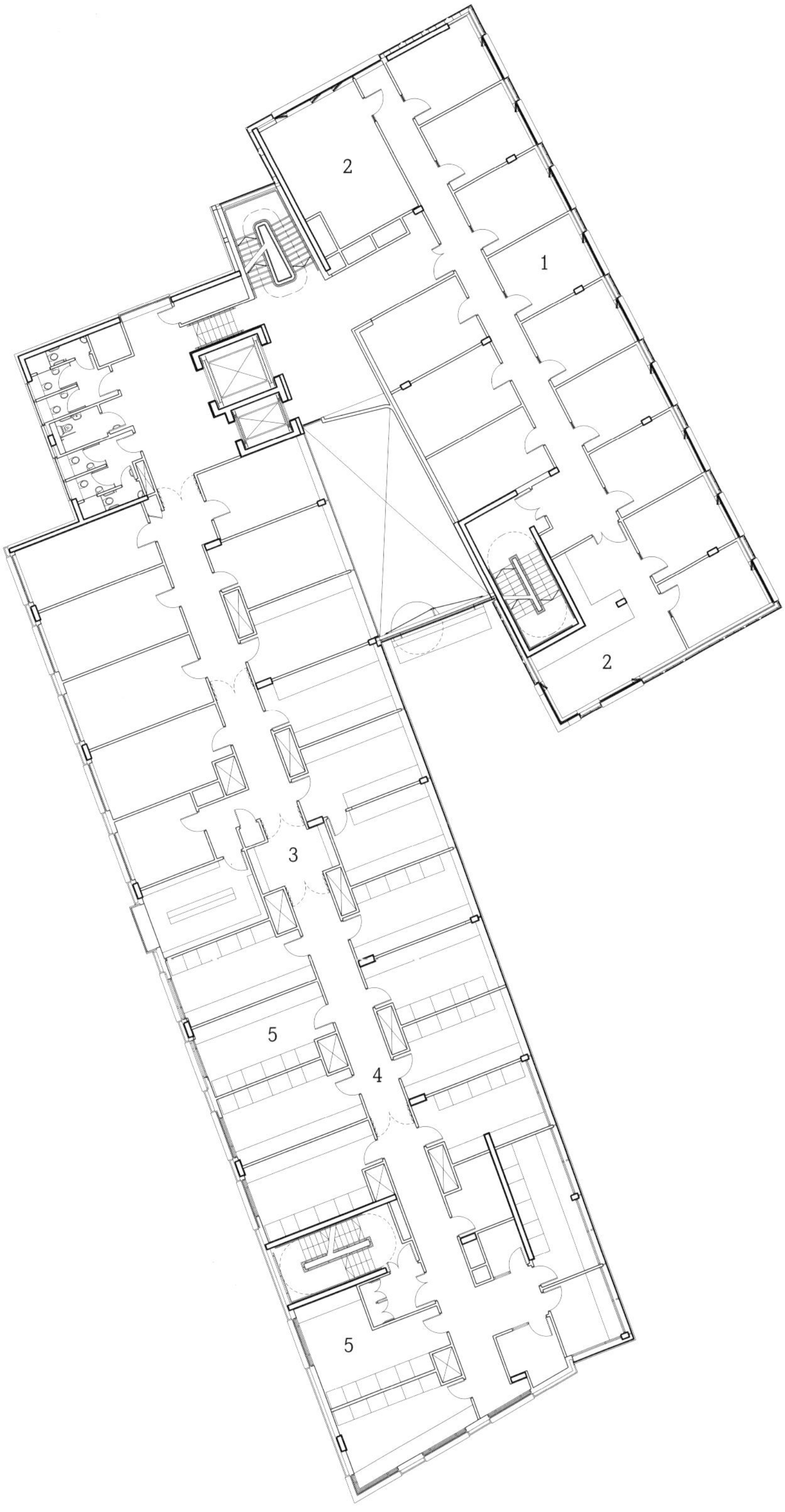

Level Three plan

1/ academic offices
2/ graduate student office
3/ gowning area
4/ clean corridor
5/ metal free laboratory suite

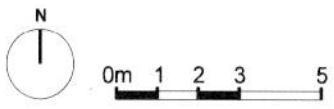

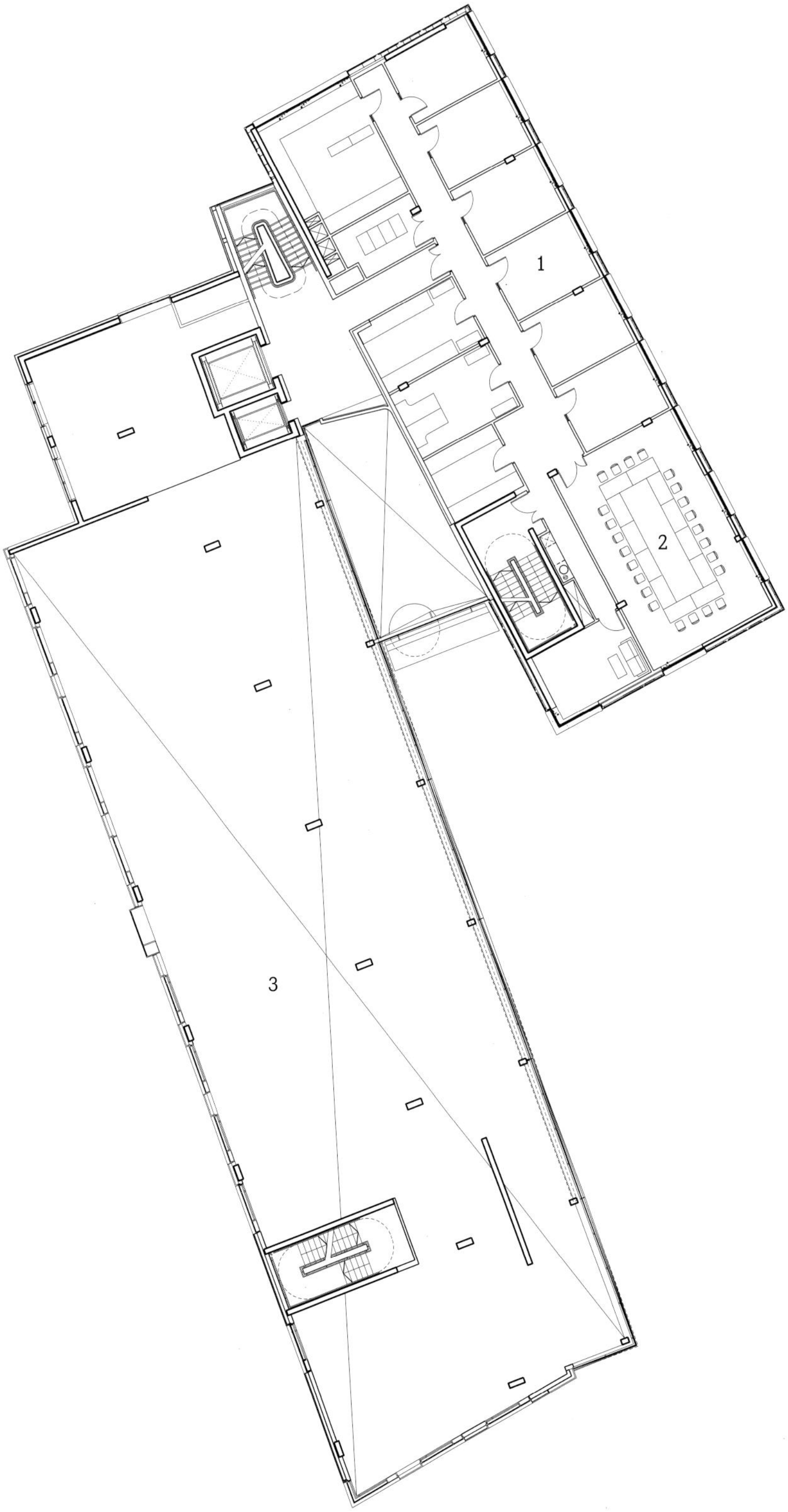

Level Four plan

1/ academic offices

2/ meeting room

3/ change of level–only four floors to laboratory wing

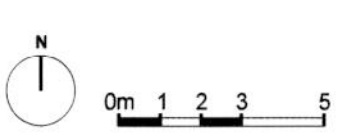

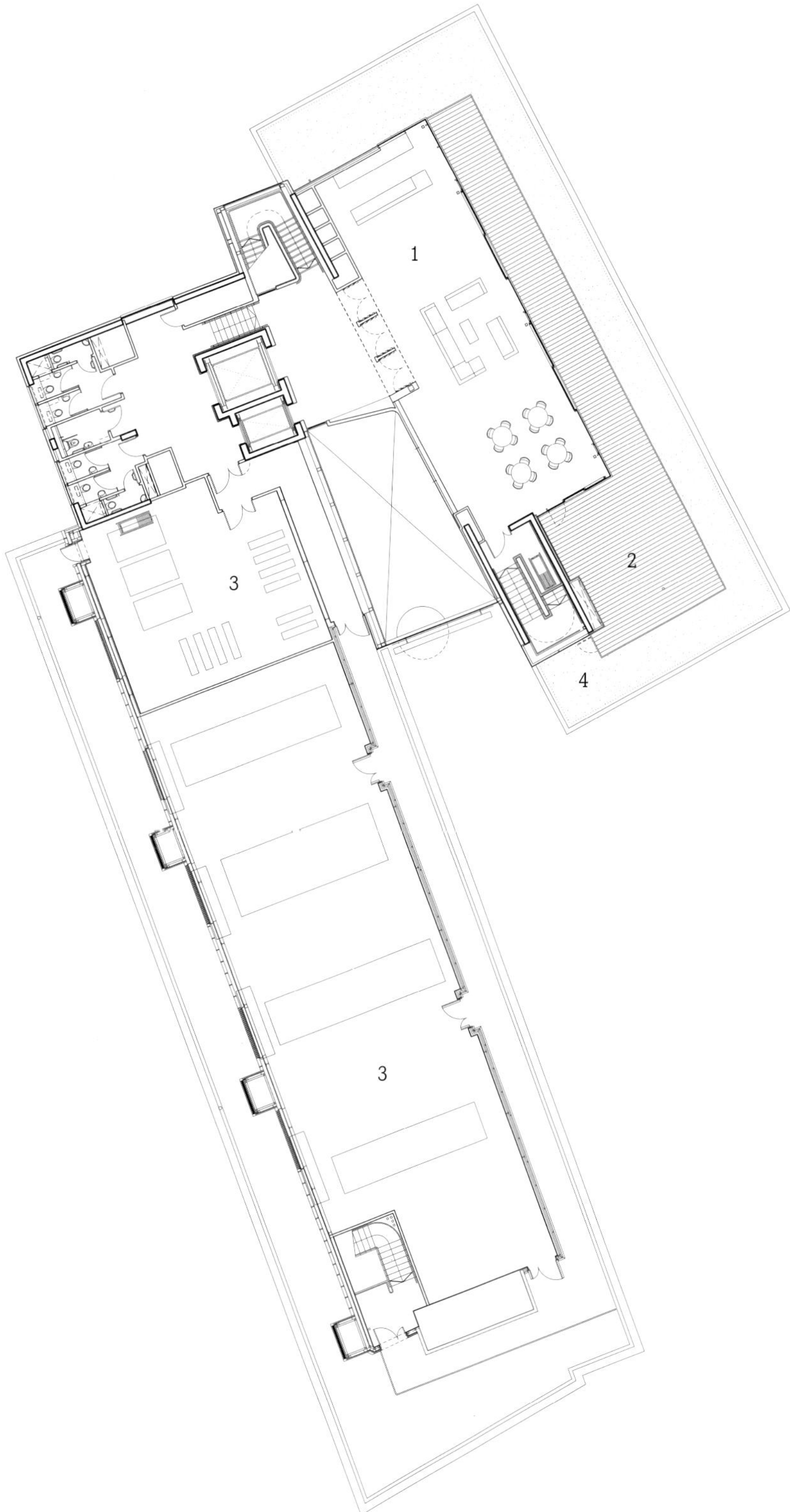

Level Five plan

1/ senior common room
2/ external terrace
3/ plant
4/ green roof

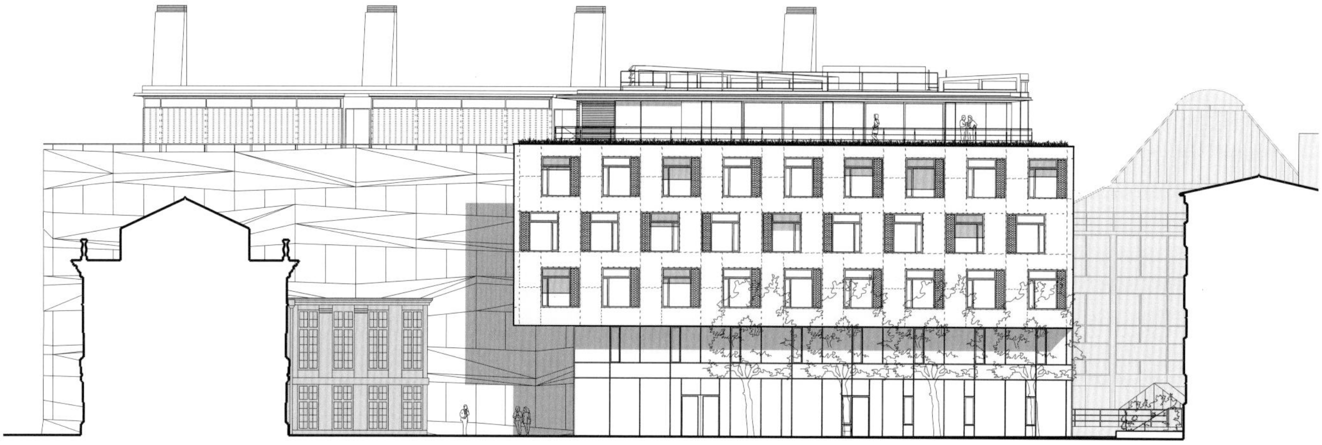

East elevation

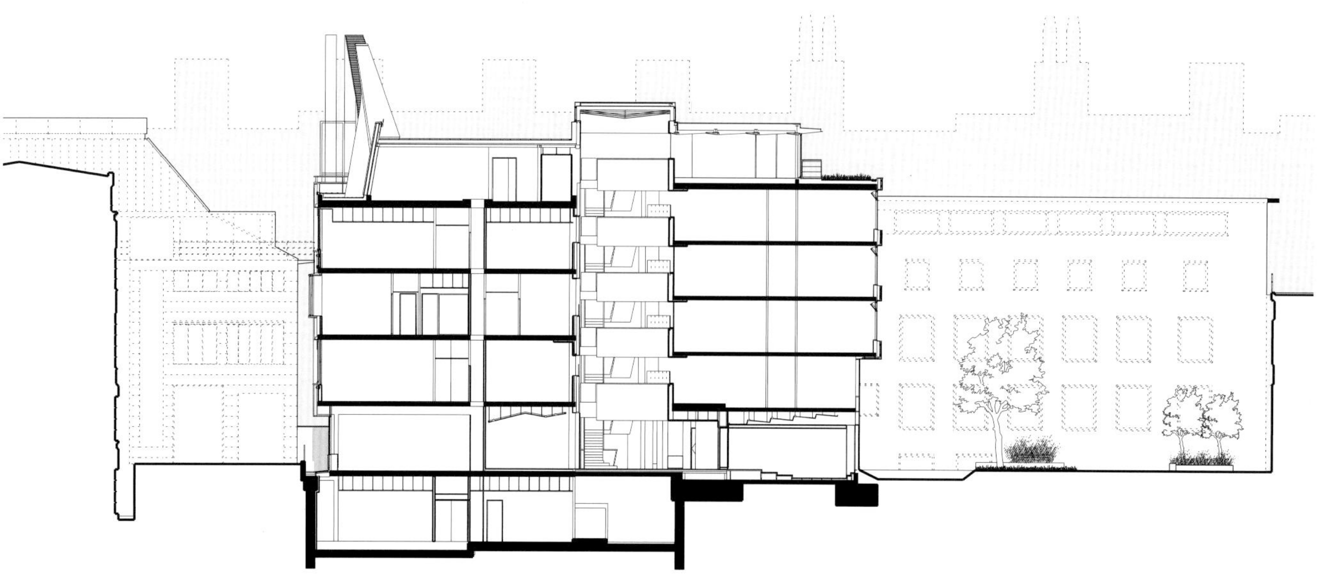

Short section

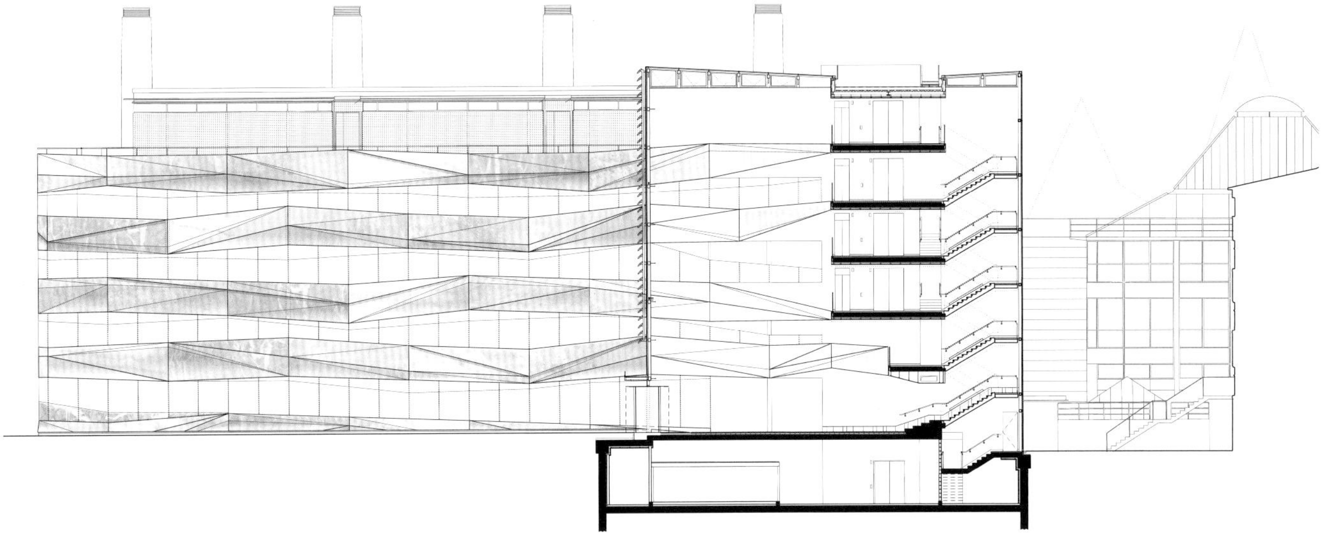

Sectional elevation from east with narrative wall

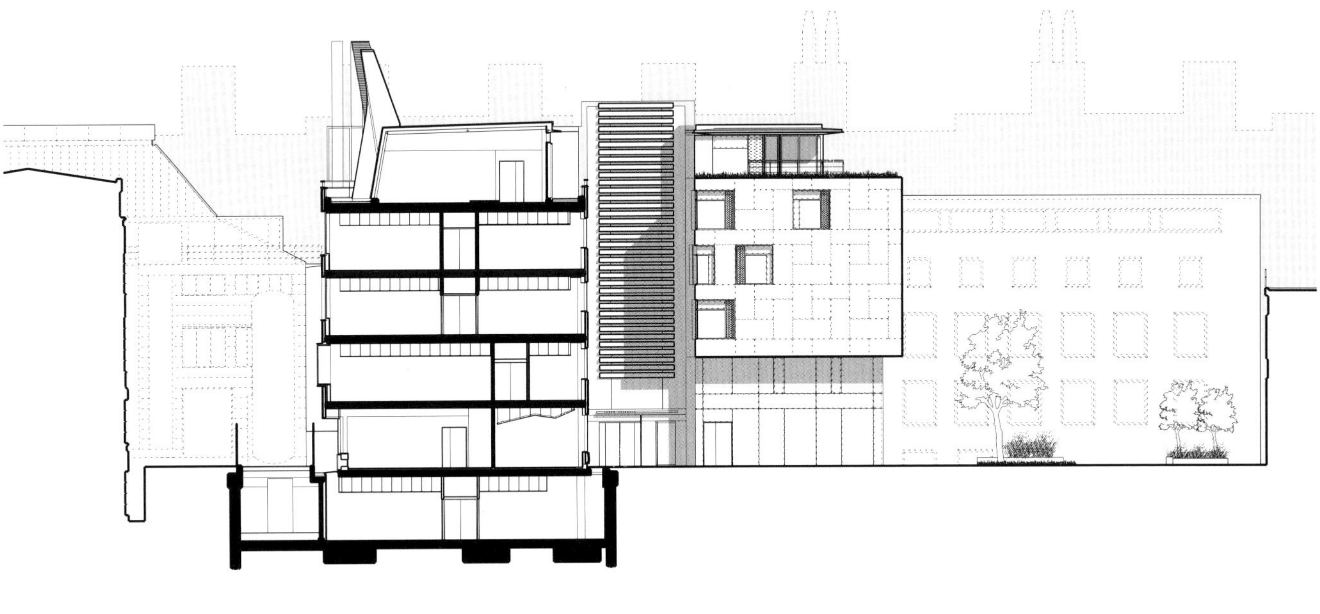

Sectional elevation from south

West elevation

South elevation

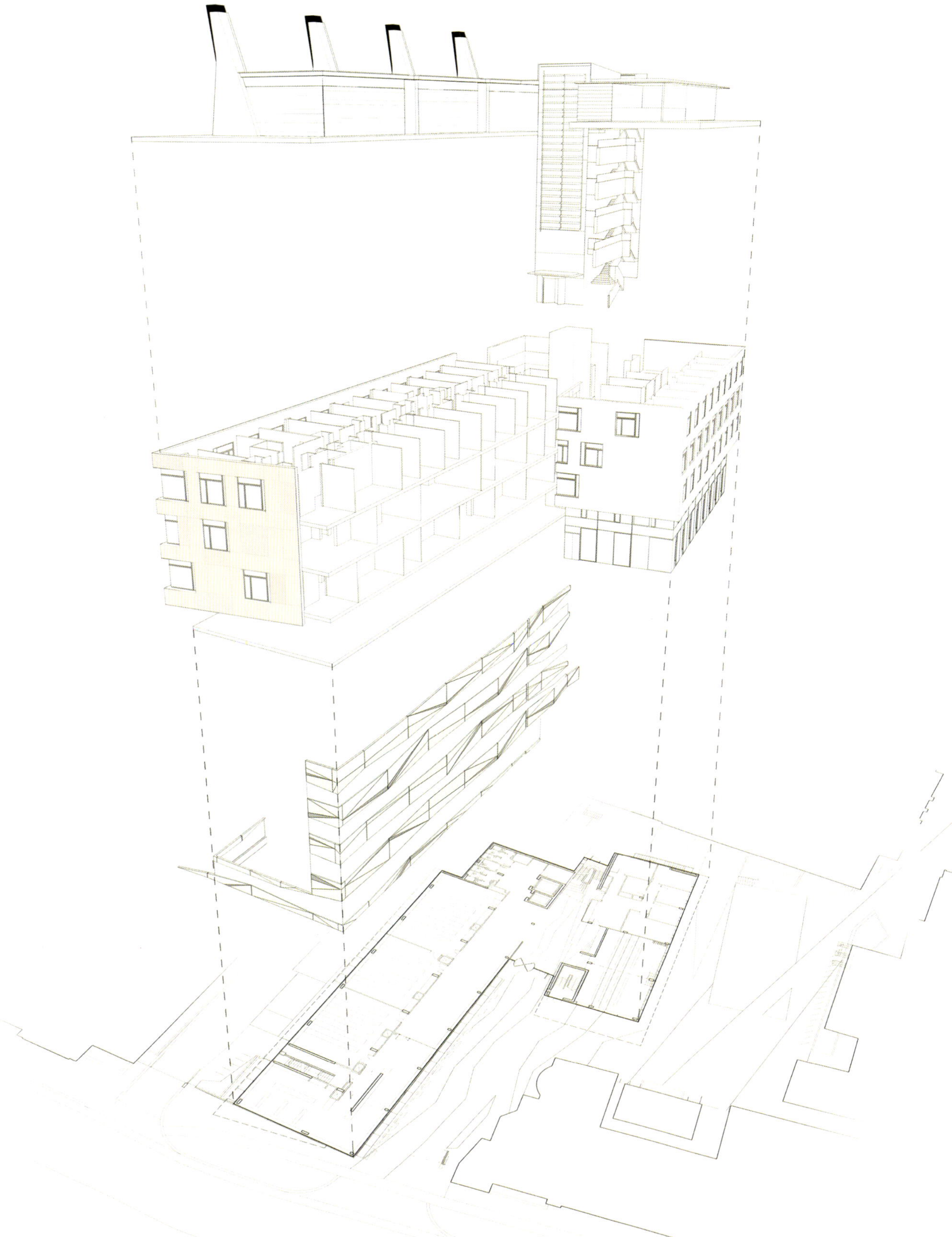

Exploded axonometric

Ian Durbin and Keith Horsley

Earth, water, fire and air

The environmental strategy and the engineering services design for the building had a longer gestation than many of its other aspects. Two years before Wilkinson Eyre first put pen to paper on the design of the new building, we were appointed to help refit parts of the Earth Sciences department's old building as a short term measure while funding for the new building was being raised. A major component of this project was the construction of a suite of metal-free laboratories on the first floor of the 1960s part of the old building.

"What's a metal-free lab?" was the first question that our engineers had to ask. Fortunately, in Professor Alex Halliday, the department had someone who had been involved in the construction of such facilities twice before, at the University of Michigan and the ETH in Zurich; and he arranged for the design team to visit Zurich to see how the challenge of servicing the labs had been met.

The work carried out in the metal-free labs generates an acidic atmosphere, and although this is largely contained within specially designed laminar flow hoods, it is impossible to achieve 100 per cent containment and any exposed metal in the lab is liable to corrode very quickly. This necessitated a different way of thinking for all members of the design team, but specifically the services engineers who were used, by default, to specifying metallic components to convey fluids, and for anything where the conduction of electricity is required. Substitutes, and sealed units, are easy enough to find for some components, but much more difficult for others.

The labs also have a cleanliness requirement, roughly equivalent to an ISO class 7 (FED-STD-209E class 10,000) clean room, although validation was not required, and so the fresh air ventilation rates are high and the ventilation systems are designed to create a pressure cascade between spaces. The high fresh air ventilation rates meant the facility was very energy-hungry. Because it was only intended to be used for three years, and was being squeezed into an existing building, the lack of space, and lack of budget, precluded certain energy-saving measures from being included, but we gained valuable experience and understanding of the issues which were later applied to the design of a similar, larger suite within the new building. The early operation of the temporary facility also raised questions about the level of resilience that would need to be built into the design for the new building.

At the other end of the spectrum from the highly serviced metal-free labs, the building contains academic offices, which it is the University's policy to naturally ventilate where possible. In between these two extremes there were a number of other types of laboratory, and the level of environmental control required varied significantly from one to the other: from Geochemistry labs with similar requirements to the metal-free suite, to teaching labs which are little more than large classrooms.

The concept of separating the highly-serviced laboratories and the less highly-serviced office and write-up accommodation into separate wings emerged early in the design process. The varying requirements of the different lab types influenced the vertical arrangement of spaces in the laboratory wing, with the metal-free labs and other Geochemistry labs on the top floor and a suite of mass spectrometer laboratories–semi-clean environments with high cooling loads–on the floor below. This strategy of locating the more highly serviced laboratories closer to the rooftop plant room reduced the length of duct and pipe runs and reduced the space required for vertical services distribution on the lower floors.

Meanwhile the academic offices and write-up areas are located on the other side of the atrium, set back from the busy South Parks Road, to avoid traffic noise intruding through open windows and allowing the perimeter spaces to be naturally ventilated. A simple mechanical displacement ventilation system is provided to serve the internal open plan areas. These internal areas are open to the atrium, which provides the extract air path.

For the metal-free labs, our team worked closely with the end users to develop the constant volume, always-on ventilation solution of the temporary facility into a more energy-efficient variable volume system that would reduce the flow rate to a laminar flow hood when its sash was closed. The extra control dampers and sensors needed for this strategy required a deeper ceiling void than that available in the temporary facility, and made for quite a sophisticated control system, but this was justified by the savings not just in fan energy consumption on both supply and extract, but in the energy required to heat and cool the supply air too. It was also intended for the metal-free and other laboratory ventilation systems to incorporate heat

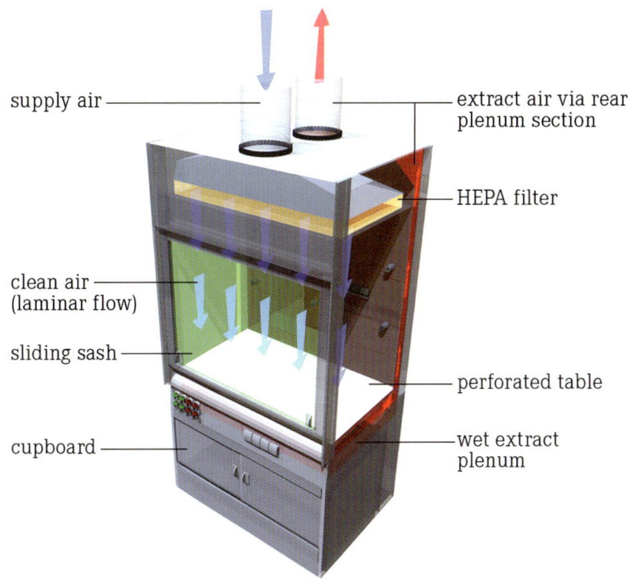

top/ Diagram showing energy-efficient laminar flow hood used in metal-free labs.
bottom/ Cross-section showing strategy to split the building into a heavily serviced laboratory wing and a more lightly serviced, naturally ventilated office wing.

recovery between the extract and supply air. Conventional heat recovery devices could not be used because their metallic components were incompatible with the corrosive extract fumes. We had successfully used polypropylene run-around coils on other laboratory projects, but these take up quite a lot of space. The requirement for more generous plant room space could not be met within the planning constraints on the massing of the building, so unfortunately this energy-saving feature had to be discarded. A proposal to provide fresh air free cooling to the IT room went the same way due to budgetary constraints, but the final design retained many energy-efficient features such as heat recovery from the IT room and non-lab ventilation systems, the fabric design of the office wing, highly efficient boilers and chillers, and variable volume heating and cooling systems.

One of our early tasks was to develop an energy strategy for the building that would comply with Oxford City Council's planning requirements and meet the client's own aspirations for carbon emissions reduction. The Council's requirement was that 20 per cent of the development's energy use must be met from renewable sources. This was a subtly different requirement to the minimisation of carbon emissions, which the department and the Estates Directorate were more focussed on. The council's definition of "renewable energy" includes heat pumps and these were evaluated along with solar thermal, biomass, wind energy and photovoltaics as possible solutions, using a crude energy model generated from an early area schedule. Although some consideration was given to a biofuel boiler running on recycled cooking oil collected from the University's many catering

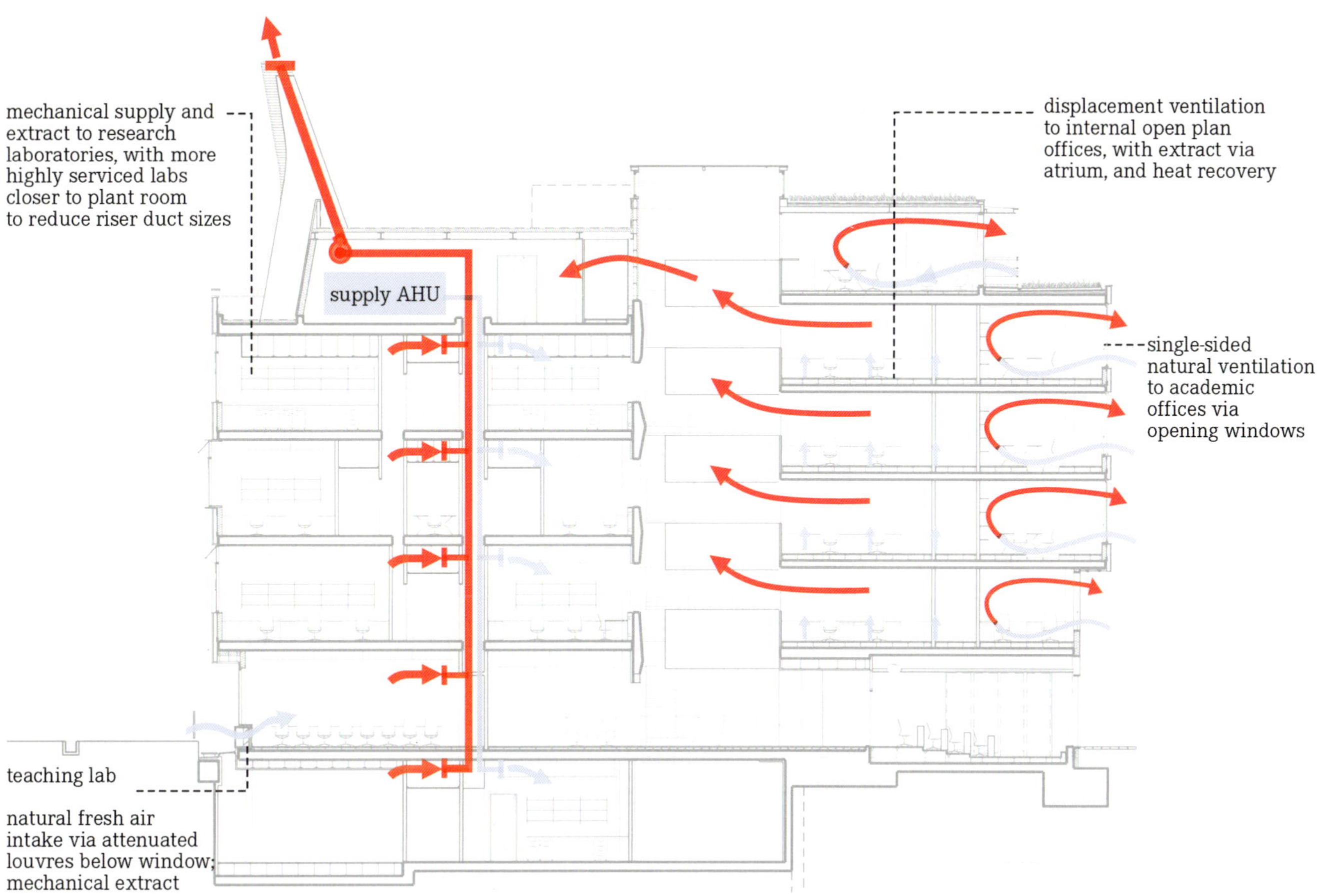

facilities, there was concern about the track record of this still emerging technology. The other contenders ruled themselves out through high capital cost, lack of available resource, lack of a suitable load, or a combination of these factors; and ground source heat pumps were identified as the most appropriate means of meeting the statutory requirements.

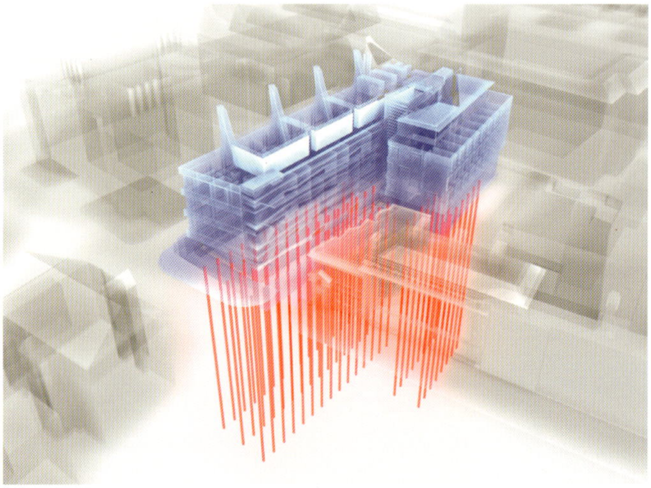

A heat pump is a device for transferring heat from a low temperature "source" to a higher temperature "sink". It works on the same principle as a refrigerator, and can be operated in reverse to produce cooling. In the Earth Sciences building, the heat pumps produce hot water for heating at around 45°C, or chilled water for cooling at 6°C and the efficiency of the process increases the smaller difference between the source and sink temperatures. Heat is transferred between the heat pumps and ground by a water/glycol solution running in a closed system through a network of vertical boreholes. The ground, at depth, is at a constant temperature all year round (around 14°C in Oxford) and therefore a ground source heat pump operates more efficiently in winter heating mode than an air source heat pump, and more efficiently in summer cooling mode than an air-cooled chiller.

It was a happy accident that the use of the ground in this way harmonised so well with the academic interests of the department. A ground source energy system also had the potential to provide significantly more than 20 per cent of the building's energy requirement, and while the project's funding initially only ran to a system sized to meet the planning requirement, the then head of Earth Sciences, Professor Philip England and the head of capital projects at the University Estates Directorate, Mike Wigg, worked to secure additional funding so that the carbon emissions reductions from the ground source energy system could be maximised.

We had a solid track record in integrating this type of low carbon technology into buildings, starting with one of the first such systems in the UK, at the Royal Veterinary College in North London; and including one of the largest ground source energy systems in Europe at One New Change in London, which was on site at the same time as the Earth Sciences building. Geothermal International was the specialist appointed to carry out the detailed

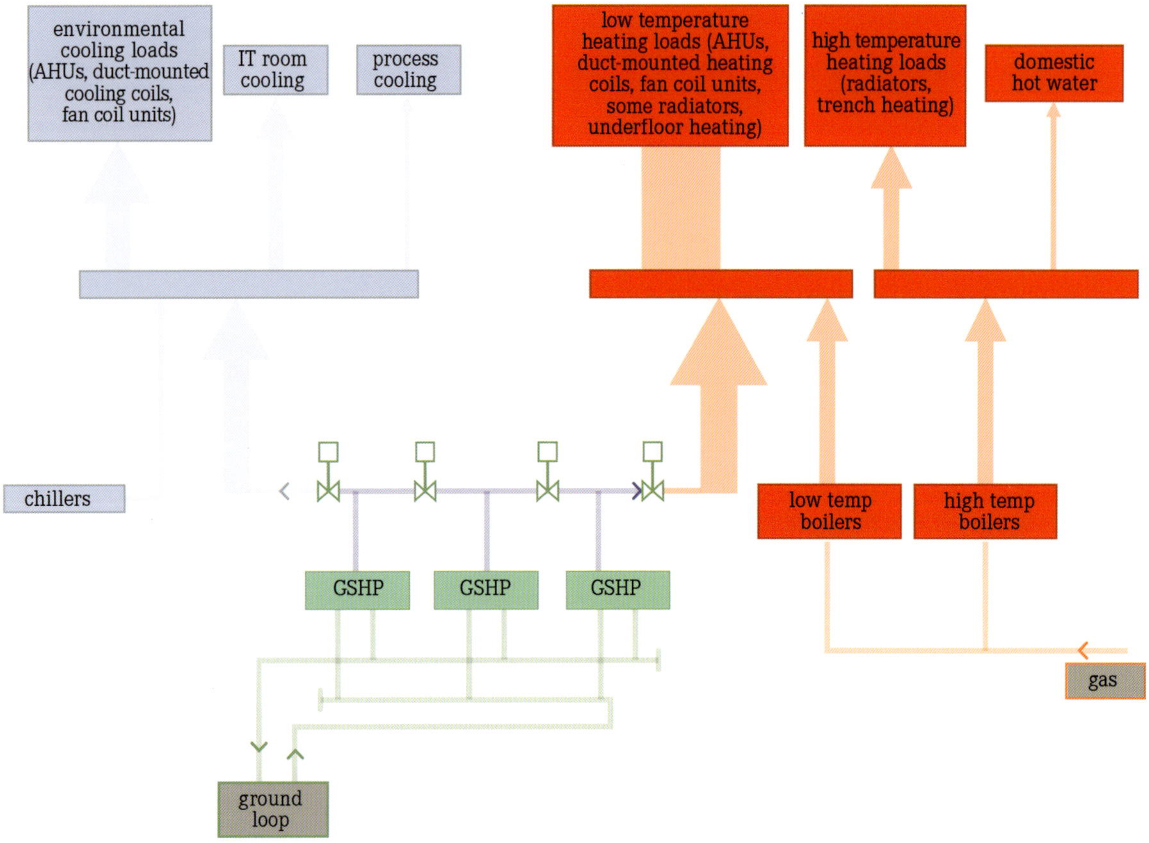

design and installation of the system. The site contained very limited space outside the footprint of the building; a carefully planned enabling works contract of underground services diversions was necessary to free up even this space. So the only place that heat exchange with the ground could take place was underneath the building itself. It became clear from the Environment Agency's comments on the planning application that there could also be a restriction on the depth to which any vertical heat exchange boreholes could be drilled. The Agency advised that the Great Oolite aquifer crossed the site at a depth of somewhere between 60 to 120 metres and they would not permit the boreholes to penetrate it for fear of causing cross-contamination between different levels of the aquifer. A test borehole drilled in advance of the main works located the aquifer at around 65 metres below ground, knocking around 40 per cent off the potential capacity of the system based on more typical 100 metre deep boreholes.

The original proposal was based on a combination of energy piles (heat exchange pipework embedded in the concrete pile foundations) and dedicated closed loop boreholes. But when Geothermal International and Laing O'Rourke met to programme the installation, it became clear that the installation of horizontal pipework to connect all the energy piles together was a significant critical path item, and the scheme was designed to use dedicated boreholes only, which was a more time-efficient solution.

At the same time as these design developments, we produced a more sophisticated energy model of the building using IES dynamic simulation software. In conjunction with Geothermal International's modelling of the borehole field, this enabled a number of options to be considered, including incorporating solar thermal or CHP into the ground source energy system to replenish the heat extracted from the ground to heat the building. Although the building's load profile is heating dominated, it was concluded that there was sufficient cooling load to make such measures unnecessary. The configuration of the system also allows instantaneous heat recovery between different parts of the building, for example using the heat rejected by the IT room cooling system to heat other areas in winter.

The heat pumps' heating efficiency decreases, and their cooling efficiency increases, as the water temperature in the ground loop decreases. The controls of the ground source energy system are linked in to those of the building's heating and cooling systems and the changeover from heating to cooling is programmed to take place at a ground loop temperature optimised to achieve maximum carbon emission savings. Extensive heat and electricity metering is provided throughout the system, and the carbon emissions saved by the system can be monitored in real time on the Building Management System. In the four years that the building was in design and construction, the quest for low carbon emissions and running costs has intensified and the University is pushing laboratory design further down the low carbon route. We are now embarking on the next stage in this journey, the new Chemistry Oxford research laboratory project, which will contain over 200 fume cupboards. All of the solutions adopted on Earth Sciences, as well as any post-occupancy feedback, will be incorporated. In addition, heat recovery will be provided on the fume extract systems, and on a number of other process systems, in an effort to achieve the aim of constructing the "greenest" Chemistry building in the world. In the meantime, the Earth Sciences building remains an important milestone in the journey towards low carbon laboratory buildings.

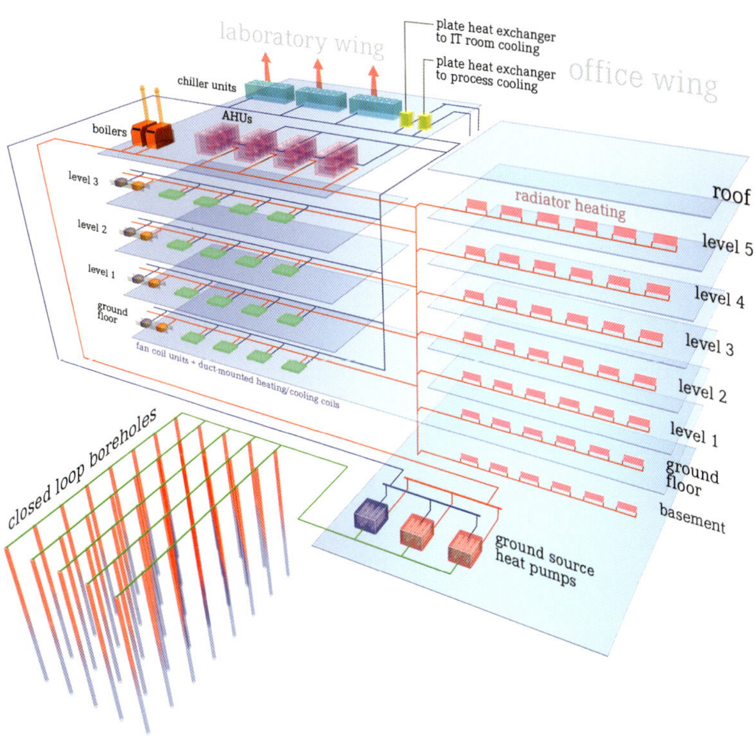

opposite top/ Visualisation showing building with boreholes for ground source energy system below.
opposite bottom/ Diagram comparing the amount of heating and cooling achieved by the ground source heat pump with energy from other sources.
left/ Schematic diagram showing the heating and cooling systems to each wing of the building.

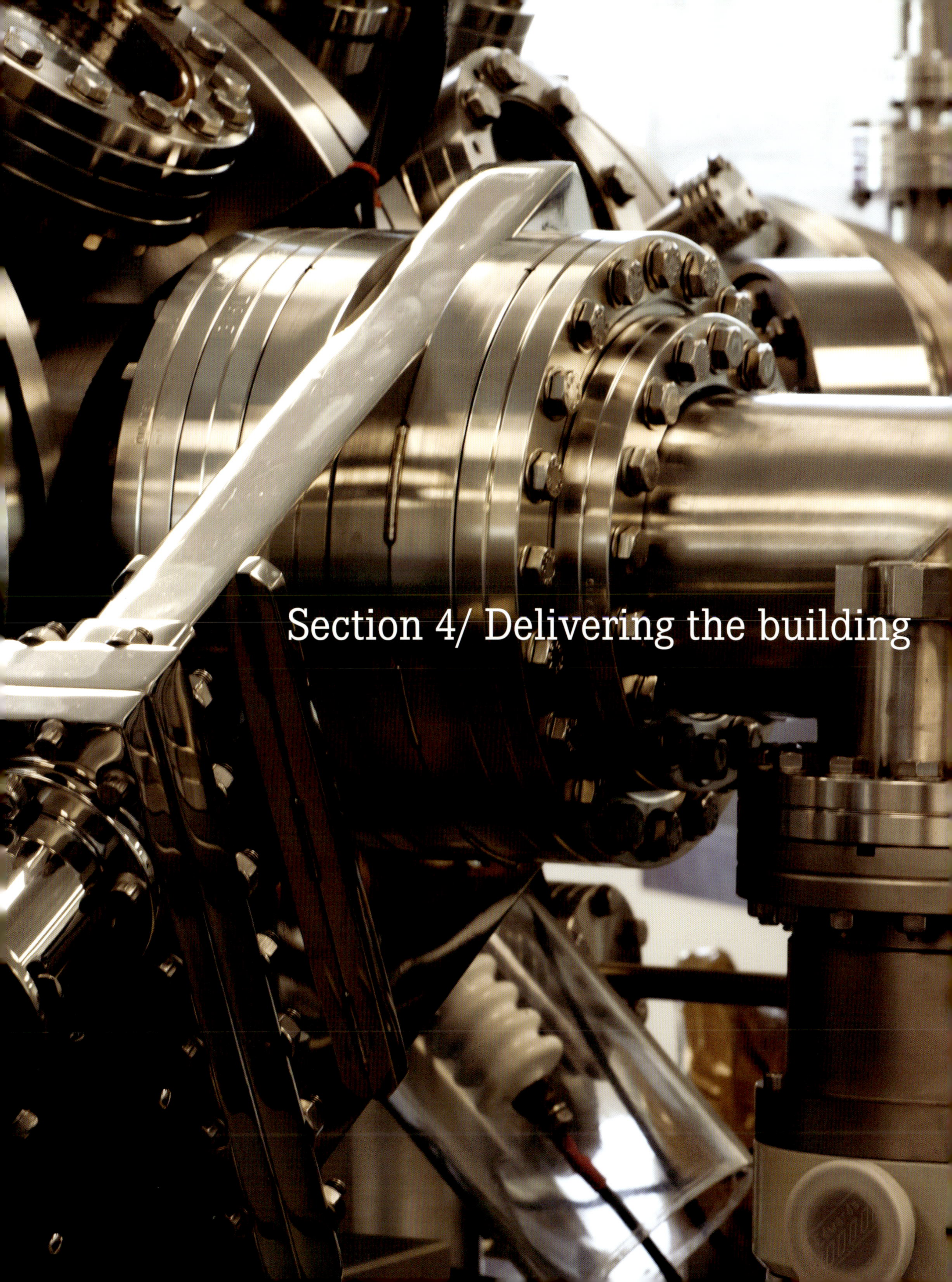

Section 4/ Delivering the building

Mike Wigg

Notes from a conversation

opposite/ The 1935 Lanchester and Lodge masterplan for the Science Area proposed three strips of development—for medical and biological, museum and chemistry subjects.

When did you first start thinking about a new building for Earth Sciences?

The University were granted funding from HEFCE specifically for upgrades to the scientific research infrastructure and several million pounds of this were allocated to the Department of Earth Sciences with a view to reconfiguring the existing building.

We appointed an architect—Berman Guedes Stretton—to develop a couple of options and, subsequently, a scheme which addressed some of the department's immediate issues. It became clear that refurbishment would be quite an expensive option and carried with it an element of risk in terms of what English Heritage would find acceptable because the existing building was adjacent to the Grade I listed Museum of Natural History.

We began to question whether trying to fit Earth Sciences' growing need for laboratory space into a building that had never been designed for modern scientific research was the right approach. It was on this basis that we began to consider the implications of a new-build rather than a refurbishment, and whether the University would feel comfortable with allocating additional funds to meet this need. They agreed to go ahead on the proviso that the department would go out and fundraise to supplement the funds available through SRIF (Science Research Investment Fund).

This is a big building for a relatively small department. Was the need to raise the profile of the department a consideration in giving the project the go-ahead?

We had a major appointment at the same time that we were considering a new-build project. Alex Halliday joined the department and it was essential that the building provided him with facilities comparable with his state-of-the-art metal free laboratory in Zurich. Part of the funding was set aside for a mass spectrometer for him to continue his research on, and it simply wouldn't have been possible to accommodate this in the existing building. To keep things moving in the short term, we did actually fit out a metal free lab in the existing building but it wasn't straightforward at all.

So that was one of the driving forces—the profile of the academics employed by the department. Alex was going to bring with him a growing support team to build the academic profile of the department, and we needed to accommodate them.

Would you say that science projects like this come forward based on academic and research priorities, then, rather than the phasing set out in the Farrells masterplan for the development of the Science Area?

Divisional priorities drive our decision making, so our estates strategy sets out a high level vision which is intended to meet academic needs and maintain Oxford's position in the hierarchy of universities around the world. It has to be fluid enough to respond to the changing needs and parameters of each Division. The masterplan developed by Farrells for the Science Area happened at about the same time as we first started discussing

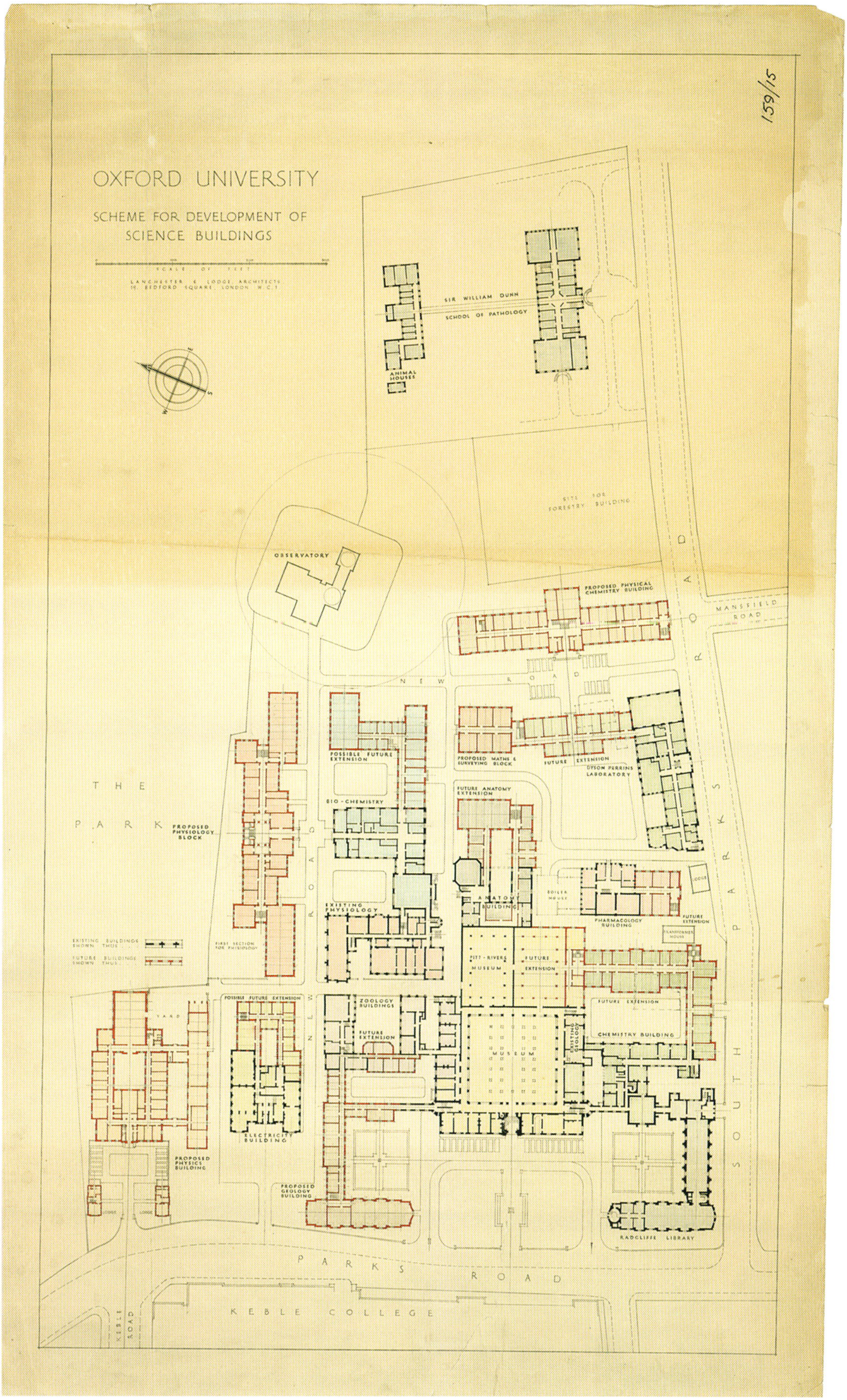

OXFORD UNIVERSITY
SCHEME FOR DEVELOPMENT OF
SCIENCE BUILDINGS
SCALE OF FEET
LANCHESTER & LODGE ARCHITECTS
19 BEDFORD SQUARE LONDON W.C.1
159/15
SIR WILLIAM DUNN
SCHOOL OF PATHOLOGY
ANIMAL HOUSES
SITE FOR FORESTRY BUILDING
OBSERVATORY
PROPOSED PHYSICAL CHEMISTRY BUILDING
MANSFIELD ROAD
NEW ROAD
PROPOSED MATHS & SURVEYING BLOCK
FUTURE EXTENSION
DYSON PERRINS LABORATORY
POSSIBLE FUTURE EXTENSION
BIO-CHEMISTRY
FUTURE ANATOMY EXTENSION
THE PARK
PROPOSED PHYSIOLOGY BLOCK
EXISTING PHYSIOLOGY
ANATOMY BUILDING
BOILER HOUSE
LODGE
PHARMACOLOGY BUILDING
FUTURE EXTENSION
EXISTING BUILDINGS SHOWN THUS
FUTURE BUILDINGS SHOWN THUS
FIRST SECTION FOR PHYSIOLOGY
PITT-RIVERS MUSEUM
FUTURE EXTENSION
POSSIBLE FUTURE EXTENSION
YARD
ZOOLOGY BUILDINGS
FUTURE EXTENSION
MUSEUM
FUTURE EXTENSION
CHEMISTRY BUILDING
ELECTRICITY BUILDING
PROPOSED PHYSICS BUILDING
PROPOSED GEOLOGY BUILDING
RADCLIFFE LIBRARY
SOUTH PARKS ROAD
PARKS ROAD
KEBLE COLLEGE
KEBLE ROAD

Oxford University - South Parks & Keble Triangle
Landscape Masterplan
1:1000@A1

St Anne's College
Woodstock Road
University Park
Banbury Road
Keble Road
St Giles
Blackhall Road
Little Clarendon Street
South Parks Road
St Cross Road
New College Sport Ground
Central Ward Sports Ground
Mansfield College
Mansfield Road
Wadham College
St John's College
St Cross College
Saville Road
Balliol College Sports Ground

which approach we should take for Earth Sciences. The new Biochemistry building was one of the early new-build elements within it. The initial masterplanning work analysed the buildings that were questionable in terms of function and condition, and enabled us to identify those which needed to be replaced. Farrells then placed a masterplan structure around those plots. We weren't really able to change the plot layout too much, but what we did want to change was how people moved round the site. We wanted to provide better clarity as to where people were in terms of the landscaping, and separate the movements of vehicles and pedestrians. Providing a proper spine down the length of the Science Area was also one of the main drivers behind the masterplan. Fitting in around this were a number of redevelopment plots that would become available, and of course once they do become available, there are important issues of decant, moving people out of buildings so that you can knock them down and rebuild them. That's the sort of juggling we have to do.

So with Alex Halliday's appointment, Earth Sciences came to the top of the pile. Did you do a lot of briefing with the department?

It's been ongoing. Initially it was the in-house project management team that were working with the department. One thing we're adamant about is that any end user department has one point of contact to interface with us, someone who can get a consensus as to what the answers to the briefing questions are. There are user groups that you have specific interfaces with, but really it has to be one person. With Earth Sciences, the department didn't have one individual that they could make available long enough, as it's pretty much a full-time job, that of departmental champion. As a result we proposed that an external consultant help with this, acting as the interface between us, the professional team and the end users to understand the department's operational requirements. Wilkinson Eyre synthesised the information into a developed brief, which was then interpreted as a feasibility study.

So we do an initial outline brief as to what the overall aspiration is, but it's really the feasibility report that is a full expression of the brief. And beyond that you go through the various design stages, and what's important then is that the developing design is signed off by the Project Sponsor Group at those gateway points.

Considered within the overall Oxford estate, is the Earth Sciences building a significant project? What is its character?

It's a strange situation because the department didn't have an especially high funding profile. But they say that the building has made a big difference. I think it's helped make that step change, and that's an important story to tell. That was one of the objectives of the project, to become one of the most pre-eminent Earth Sciences department in the world, and now they've got some very important academics working in the building and hope to continue to grow their expertise and their profile.

Do you think the same scenario would happen today? Has the character of the University changed in scrutinising major capital projects?

opposite top/ Farrells' 2007 masterplan for the Science Area set out the principles for future development, identifying buildings to be retained, zones and individual plots for redevelopment, and improved landscaping, access and pedestrian circulation across the site.
opposite bottom/ The masterplan as built/proposed in 2011.

The University's expectation is that the Estates Directorate will get them value for money, so one of our objectives is to ensure that the brief is not gilded at all, that there is a definite need, and that each project can be justified. The Division or individual department have to produce a business case which includes how they're going to fund the building and which reflects on the running cost. We have a space charge which calculates all the costs per square metre when the building is in occupation and obviously lab spaces are more expensive. This all goes into the business case, which is then scrutinised through the University's committee approval process.

Here you effectively have the Science Area in the middle of the city. Do you think you could be putting up these buildings a year quicker without the planning constraints?

Obviously the building process is more difficult on a constrained city centre site. But the academics like the interface with the colleges. Most say that they prefer to be based centrally in Oxford. There are certain departments and units that can operate in the peripheral sites we're developing outside the city, but ongoing development in the city centre is inevitable.

How do you choose the right team to deliver each project?

When I came here six or seven years ago there was no consistent process in place for delivering projects. We've now got a comprehensive methodology set up, with a process map, ongoing monitoring and so on. One of the key things within this structure is supply chain management. We realised that we needed to establish some form of informal framework arrangement. For other areas we have a full framework in place which we use as a way of shortlisting suppliers who can then tender for individual appointments.

One thing we're looking to do better is to set up some Key Performance Indicators (KPIs) so we can identify how various inputs are performing and use that as an incentive to continue to improve. Otherwise suppliers might drop off our preferred list. That goes for contractors too. We did think of having a framework but we ducked away from it in today's market.

For the new Biochemistry building we did quite an extensive prequalification and shortlisting process for the main contractors, and what dropped out of the early tenders is an informal framework group who tend to be the ones we tender to.

We did undertake another prequalification for the Maths building at the Radcliffe Infirmary Site and similar names came out in the shortlisting because they had made a real effort to understand the challenges of working in Oxford and in particular how they would work with the academics within the University. An ability to understand the elements that can influence a contractor's ability to perform and deliver is what we look for.

We were very keen to improve the quality of architecture that the estate was delivering. We looked around and there were some practices which seemed to be producing good buildings, and we wanted to encourage them to think about working at Oxford.

You obviously have to set out certain parameters when you are commissioning architects, but with Earth Sciences were you confident that you were getting the right fit with Wilkinson Eyre?

That's very important and we didn't jump in quickly. It's fair to say it took a trip to visit the National Waterfront Museum in Swansea, and sitting and talking with the team on the train to do it. It was an interpersonal thing rather than credentials of what had been delivered previously. The fact that Stafford Critchlow, the Project Director, was really sensitive to the culture of the department, and understood what their requirements were that persuaded us. Stafford then went with the department on a field trip to better understand their research interests. You can't underestimate these things. The personality mix has got to work. We're great believers that team spirit is critical to the success of a major project.

So rather than a competition process over a couple of months, you've evolved a way of doing a much quicker competitive interview. It seems good in a way that you don't do a formal competition because picking a scheme which isn't tested and costed can lead to disappointment on the part of the end users.

We've tried various approaches and I think what we like to do, and what we did in Wilkinson Eyre's case, was to give the team a flavour of what we wanted to achieve, and they came back to us with their initial thoughts. They indicated how they thought the site issues and challenges could be met, but then we went back to the beginning again once they were appointed and started with a proper briefing, working from the inside out. And that's important, that these buildings are designed from the inside out rather than us asking how we fit everything into a wonderful vision. That's the way we like to do it because when it comes to the design it has to be the result of a good relationship between the architect and the client body.

And that's an important distinction between you and other universities. Sometimes the client gives a very detailed brief, virtually the room numbers....

Yes–we're choosing an architect, not a design. That's the most important thing.

The other aspect of the way the building was delivered is the procurement route, the fact that it's design and build.

Yes, I feel strongly about making the right choices regarding the procurement route. There's quite a bit of debate about the pros and cons of two-stage procurement and a lot of cost consultants say that you get better value using this approach. I think some people don't realise that two-stage tendering is fully competitive and gives you the opportunity to bring supply chain input upstream–which in turn should give you a better outturn result.

You have to be prepared to get rid of the contractor you appointed in the first stage if the commercial negotiations don't develop appropriately. But the biggest incentive for everyone to perform is the ongoing workload. We've been spending £100 million a year for the last few years and that's not insignificant. It's one that the main contractors want to be part of. That's the carrot we use and the stick is there if we need to use it. Again, it's bringing the whole supply chain upstream and solving the problems with all the professional input you can get hold of.

Who has the most influence in terms of aesthetic issues? Do you guide the academics in issues of architectural style?

I think every element of the design is subject to critique. It's only fair that the academics and end users are involved in the aesthetic of the building, and certainly most of them love to get involved in the design. They love to critique things: that's their very nature. They are researchers and thinkers, and they interrogate and question every decision that's taken. It's healthy to do things that way, and means that issues too often taken for granted get revisited and discussed. It's good for the design process.

How about the longer term aspects of a build? How much you have to consider issues of maintenance and so on?

There are a number of things. The long term sustainability for example. You try to make buildings as flexible as possible so if, say, a department declines in numbers or its research needs change dramatically you can find alternative uses for the accommodation. We also have a strong environmental sustainability agenda which is supported within the University at the highest level. And since we involve the facilities management team at an early stage in the process, we ensure that ease of maintenance is taken into account in the design, and that cost in use is kept to a minimum.

If buildings have got to work better for longer, suddenly it becomes much more important how they fit into the cityscape.

I think in Oxford the planners are always going to have that long term vision anyway because of the sheer quality of the existing buildings within the city. It's an interesting aspect of our ongoing development of the Science Area. The planning department think of the Earth Sciences building as landmark, as are the Biochemistry building and the Chemistry Laboratory II. But we're conscious that all of the buildings we develop need to be respectful of their environment. Some will need to be more subtle background pieces rather than dramatic statements, reflecting the historic buildings around them. In everything we build there has to be a strong level of awareness of the overall cityscape.

Aerial view of the Science Area.

Mike Morris

Realising the project

My philosophy is that you are only ever remembered by your end users at the end of the day, and that is something I have discovered over the years, having built Chemistry I, then Biochemistry and now the Earth Sciences building in Oxford. These commissions have taught me that the key to the delivery of any project is understanding who the real client is. Although our contractual client is the Estates Directorate, our ultimate clients were the end users of the building.

With every project we do at Oxford, we are improving our understanding of the scientific research that is going on here. Compared to when we first started working with the University, we now feel confident that we know what the client is talking about when it comes to how the labs work and so on. Once you break down the brief you can begin to understand what their requirements are. Luckily we have got a great team now. It has grown over the last eight to ten years we have been working here so there is a lot of experience between us. People say you get complacent but we are always challenging ourselves, and we are not getting complacent.

Obviously academics are an incredibly intelligent group as clients. Scientists tend to be quite off the wall, but once you get to know them it's great. We have been able to tap into their knowledge to help us, particularly on the Earth Sciences building, getting experts such as Steve Hesselbo to help us out with the stone selection, for example, and of course interrogating our ideas for the building's ground source heat system. We were up against two or three other teams at interview, but we won the job with

our track record working on the Chemistry and Biochemistry buildings. The logistics of the site were an important consideration right from the beginning. During the tender period, we spent a long time thinking about how we were actually going to deliver the job on that postage stamp of a site. Like Biochemistry, it was surrounded by existing buildings, but here there was even more of a challenge because it is on a main road. At least inside the Science Area we had some control over the roads, but South Parks Road is a fairly important artery through the city. We specifically selected a team who had done similar jobs in the past, and who understood how things work in Oxford. It was all down to planning and delivery, and people understanding that there was nowhere to store materials on site, that we had nowhere to pull in wagons, and so on. We had holding areas outside Oxford for the suppliers so we could call them in when we were ready. They would park up and we could call them in within 20 minutes once we were ready for them on site. It was that just-in-time delivery planning that was one of the biggest challenges.

We also had to consider how we were going to deal with the neighbours. Obviously the building has a lot of them, and it was important that they were fully aware of what we were going to do. We would do tea and biscuits for them, every other Friday morning, and we would tell them what we were planning over the next week, how they would be affected, and we would arrange new routes to their stores and so on.

There was a huge amount of enabling work to be done before construction started on site. There was an existing Chemistry

opposite top and bottom/ Demolition of the existing 1960s curtain-walled Chemistry building on site. A condition of planning was that, due to the architectural interest of the curtain walling system, the building was carefully surveyed and recorded before demolition and the information archived with Oxford City Council's Planning Department.

below/ General site view showing the walls propped to allow for the excavation of the basement.

building on the site and we had a massive amount of existing services going through the basement which we had to divert onto gantries all round the site. We also had to change the major power source because we had a new substation going into the basement, diverting it while we were doing the demolition and decant. So there was quite a lot of tricky stuff before we even stuck a spade into the ground. It all took four or five months, and then we started with the main contract works. It was a massive challenge, that first period.

Earth Sciences was the first building we had done with ground source heat pumps–and the first time that the University had done one. It was a completely new avenue for us, and we had to integrate it into the construction phase of the basement. We drilled down from ground level because we did not want to put a rig down into the basement during the winter, and then when we dug out the basement they had to work around these black plastic pipes. It was very interesting. We were concerned about going into the aquifer, which we'd been told was at about 100 metres deep, but when we did a trial we found that it was only down about 65 metres in the end, which reduced the capacity of the system. It seems to have been a resounding success though.

As soon as we got involved with the project, we started looking at how much of the building could be prefabricated off-site because we were working in such a constrained space. That is why we went for the twin wall structure and the precast columns. These cores, the lift shafts and the post-tensioned slabs, were all things that we brought in at tender stage to improve buildability. They had other advantages though. The pre-tensioned slabs meant we could work more quickly, and reduce carbon overall because they did not use so much material. Using the precast in situ columns also meant we could reduce the risk of having a large team working at high levels.

I think we are at the cutting edge of what we do in terms of looking at prefabrication techniques and ways of minimising risk on site. We are always looking to the next challenge. We are always trying to reduce labour because of the health and safety risks, and a lot of that is down to using prefabrication techniques.

Another challenge was how we were going to lift the precast panels onto the narrative wall facade, because some of them were not within the reach of our crane. We needed to get a crane big enough to lift them, but also with the right reach, because obviously when you get a certain distance out on the jib of the crane you cannot lift quite so much. We had to get a mobile to do some of the panels on the corner of the lab wing in the end, which meant we had to close Robinson Close for a few days.

And of course all the time we were working within a few metres of all the nitrogen tanks on Robinson Close. They are a huge risk and they are nothing to do with the building–they feed the Inorganic Chemistry department. We had to make sure nothing would affect them. A lot of thought went into developing methodologies for how we would work around them, and how we would ensure there was no danger of things falling off the building onto them. That was one of the biggest logistical issues we had to overcome.

There was nowhere really to do anything so we used Preston platforms which bolt onto the side of the building. They are like a loading bay that you can use to get all the materials in. We had them pre-loaded so all the components were in the right order for the team to use.

This is our third scientific research building in Oxford, and critical to our success has been doing mock-ups of key areas of each building. We have done more and more as we have gone along, and it has helped us realise what the important things about the building are for the users, what cannot be compromised. With Earth Sciences, there were the laboratories, especially the metal free ones, and the feature staircase that looked as though it might be difficult. And as soon as we got access to the office wing we mocked up an office, put on a door and tested to see if the acoustics worked. The metal free labs were mocked up in the bottom of the Hans Krebs tower, just to work out all the issues we had. They had been a huge challenge. We had done plenty of generic labs with fume cupboards before, but these required a completely different approach. If you have got any metal in the room, you have got to find something to replace it. It is the environment–the acid rich air–which means you cannot have metal. So where you would normally use a metal screw, you have to find an alternative. The department had had a temporary metal free facility built a few years previously in the old building, but they had not got the details right. So we visited some laboratories at Southampton University and they were great. We saw that they had lined the walls and ceiling with vinyl, so essentially they had built an upside down room. As soon as we saw it I knew that it would be the solution for this job, because we had

opposite/ Existing services are diverted onto gantries around the site while the boreholes for the ground source heat pumps are drilled from ground level.

below/ Heading up the boreholes for the ground source heat system.

been worrying about how we would do all those junctions with normal plasterboard. We very quickly decided to go down the same route. We had to work with Hoare Lea to work out how to do the access hatches and so on, and with the subcontractors we eventually came up with a solution which we mocked up. We got some real engagement from the client when we took them in to see it. We got some great feedback which is paying dividends now. The clients are very happy with the metal free laboratories and they performed well in testing.

We had never built anything like the field free labs before either though. It was part of the brief when we went to competitive tender but at that stage I had no idea what constructing them would entail. Essentially the room is built of very thin layers of steel with around 3.5 per cent silica. Once you have aligned the steel's magnetic domains, very little of the external magnetic field makes it through because when it hits the outer layer it just gets channelled around the room. You then leave an air gap and build an inner layer which does the same. The outer layer should take away about 95 per cent of the external field, and then the inner layer will take out 95 per cent of the five per cent remaining. So only a fraction of a percent makes it through into the room, which has more or less no field. It means the researchers can measure very weak magnetic properties of natural materials because they do not have to worry about the external field.

The department showed us photographs of how they had built the previous one themselves. It was unbelievable. It was like a hobby to them, the way they had built it. They had found the steel on a dock in China. It just happened to be there in the quantities they needed so they shipped it across. We started researching where we could get a similar kind of steel and we found a place in Wales where they make turbines. They understood exactly what we wanted. But it was challenging to build. You could not go in with a drill and leave it on the floor because it would affect the magnetic field. And then we discovered there was a huge HV cable running over the top of the room which could not be diverted anywhere else. In the end we had to put the cable inside an insulated pipe to isolate it.

We are still very engaged with the supply chain that helped us to solve issues like that. It is good to get confident, professional people on board because it always pays dividends. We might be paying a bit more money up front but we're getting better results. The joinery supplier, for example, does 1:1 drawings so that we can see exactly what is going on. It is fine looking at CAD drawings, but it is so much easier when you have got a 1:1 drawing because you can actually see how things are going to work.

I am personally a great believer in understanding what the architects and clients want to get out of the building, and ensuring there's enough money available to deliver it. It would be good if we could make savings down the line, but we would never want to dumb down the design. Obviously we have had to make compromises along the way on this project–for example, with the narrative wall–but without those the building wouldn't have gone ahead. Earth Sciences has been a good job for us and I do

not think we have cut any corners in giving the client what they wanted. When the information we are given is good, for the detailing and so on, it makes things a lot more straightforward. And when the brief is strong it helps too: what we signed up to is what we have all agreed to, so there should not be any surprises.

We have learned things from the project. At a recent meeting the client was surprised by the size of the fuel bills for the building. It will settle down over time because they can now look at how the building can be fine-tuned. Maybe it was an issue with the brief, or we should have given them more zones so that it would be more straightforward to control. We could have given them different scenarios as to how they could limit the amount of fuel they were using, make it more efficient. That is a key issue going forward–making sure users understand how the building runs–because fuel is not going to get any cheaper. What we need to do as a team is to give them more control, perhaps by introducing zoning within the building, and also helping them to monitor energy use.

The whole job went very well from start to finish. Contractors sometimes feel that they are quite a long way down the food chain in the construction industry, but I remember that the department couldn't believe how quickly and efficiently we were building, and it is good to see that they thought we were quite clever in what we were doing. I think they could see an analogy between what we do and the way they plan their research in sequence so it is done in the most efficient way. It was quite nice feedback from a group of boffins, Nobel Prize winners and so on.

opposite/ Site view showing progress on office wing as basement works to the laboratory wing continue.

below/ View showing proximity of nitrogen tanks (feeding the adjacent Inorganic Chemistry department) to the site.
overleaf/ This sequence of shots taken from the site webcam on South Parks Road highlight the complexity of enabling works to the lab wing in contrast to the rapid construction of the office block.

David Sivyer

Behind the envelope

Structural engineering and what it entails is not generally known or understood by many people outside of the construction industry. Our role, putting it simply, is to design the skeleton around which the architect forms the building. We effectively design the elements that support, protect and house each of the aspects that make up a building. The skeleton, or frame, needs to be designed specifically to suit the specifications of each building and is shaped by the architectural form.

The majority of recent new buildings in Oxford have basements formed using concrete walls. These are either placed in the ground using the diaphragm walling technique or by interlocking secant piles. We wanted to move away from this approach at the new Earth Sciences building for three reasons–cost, space and water ingress. Concrete walls can hold back the ground and stop most water from entering the building but they cannot be guaranteed to exclude all water. This means that an expensive and space-hungry drained cavity would be needed. Initially the basement under the building was to be two storeys, which meant that we would have had to use a concrete solution since there was a risk of not being able to push steel piles to the required depth. However, a change of brief resulted in the requirement for only one level of basement and so we were able to look at a steel alternative. Our approach was to push steel sheet piles

into the ground and weld the joints to provide a watertight wall. This saved both cost and space as the area for the drained cavity no longer needed to be built.

The ground conditions at the site consist of a layer of made ground over terrace gravels, over Oxford Clay. The sheet piles needed to be installed through both the made ground and the gravels into the clay in order to form a stable wall and to create a seal. It would not have been possible to install the sheets by percussive methods due to the sensitivity of the adjoining buildings and occupants to noise and vibration. However, the contractors were confident that they could jack the piles into the ground. It was accepted that a minimal amount of water jetting could be used in the gravels if high resistances were met, but in the event the piles were installed with little jetting and no disruption was caused to neighbouring properties.

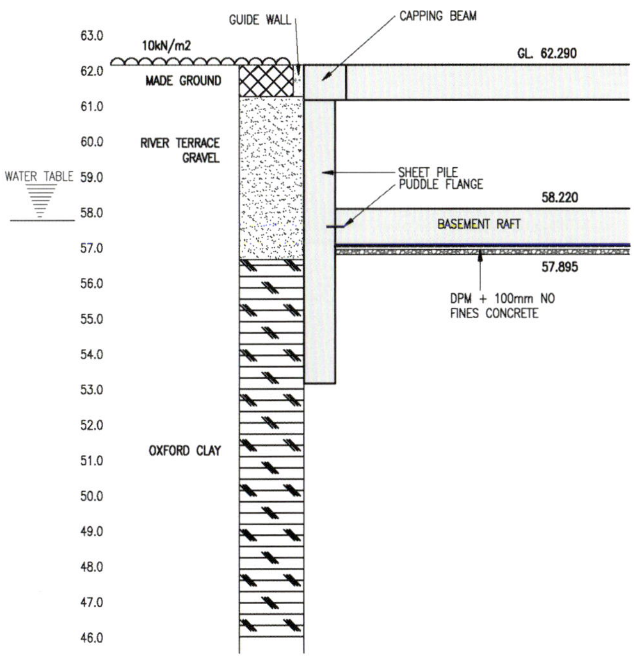

opposite top/ Cladding is applied to the skeleton of the building.
opposite bottom/ Sheet piles are installed to form the permanent retaining wall to the basement.
left/ Geological section showing piling solution.
right/ Props supporting the walls as the basement is constructed.

The wall was temporarily propped at ground level to facilitate excavation without significant deflection. The interlocking joints between the piles were welded to form a total seal against water ingress. The base slab was then sealed with a proprietary external bentonite layer together with a hydrophilic water stop against the wall.

The department's mass spectrometers and other specialist equipment had historically been located within basement spaces. Many of the technicians responsible for this equipment never had the opportunity to work above ground in a naturally lit environment. It was always the intention that working conditions for all users would be improved in the new building and we therefore looked to house equipment, wherever possible, on the upper floors, to provide a better working environment. The large 1700 mass spectrometer for the department was constructed specifically for the new building and we were given a set of specifications for the supporting floor. These set out how much movement, in the form of acceleration (displacement and time), the equipment could accommodate. The problem was that, to our knowledge, these extremely strict criteria had never been met before by a suspended slab structure. We—working in collaboration with the manufacturers of the mass spectrometer—needed to create a set of limits that the building could cost effectively accommodate. We established that a mass spectrometer works at an extremely high frequency and as such should not be affected by normal building movements—so by the time, for example, the building had deflected due to wind the machine would have taken millions of samples. It was established that to affect the readings the building would have to move so fast that it would effectively hum. It was more likely that specific incidents, such as a large sample of rock being dropped or human activity immediately adjacent to the equipment, could result in a misreading being taken. With this in mind, the final design for the building included the construction of an economical reinforced concrete frame with a standard vibration response. In order to have a certain factor of safety the mass spectrometer was designed so that it could be placed on a vibration isolated concrete plinth if necessary.

After we had completed the building we carried out a series of tests at the manufacturing facility in Wrexham. This allowed us to test likely events that could cause the mass spectrometer to fail, take false readings and determine whether the isolating plinth was required. The testing method was to perform certain actions around a working mass spectrometer set up on the shop floor of NU Instruments' manufacturing facility and then

below/ A plan reveals the building's structural grid.
opposite/ Construction of the post-tensioned concrete floor slabs.

compare these tests to recordings taken on the completed building frame. For these tests we arrived with some specialist recording equipment, a medicine ball, a project manager (weighing approximately 85 kilograms), a design manager (weighing 75 kilograms) and a structural engineer armed with a trolley with a broken wheel. With this equipment we were able to simulate the likely vibrations caused by general activities around the machine. However unscientific our tests may have looked to the highly specialised technicians at NU Instruments, they did prove that the machine was capable of operating accurately in a range of conditions, and that the building frame was suitable for housing the mass spectrometer, the isolating plinth would therefore not be required.

The building's floor plan is split, with a laboratory and an office wing situated on either side of the central atrium. Each wing had a different use and therefore had different structural and servicing requirements. Both wings were formed from a concrete flat slab but the grid and floor-to-ceiling heights were

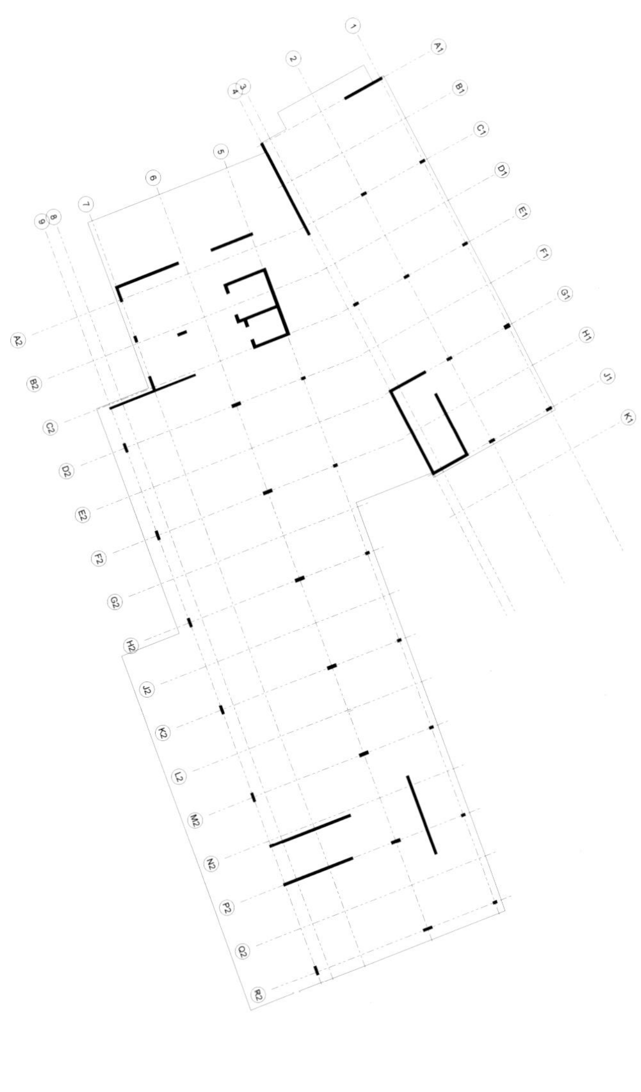

different. In order to squeeze in the additional floor to the office wing the structural slab needed to be very thin. This challenge was made more interesting by the large cantilever at the front of the wing. The typical flat slab required to achieve this would have been too thick and would have meant breaching the Carfax height restriction, while introducing beams below the slab would have impacted on the design intent and the exposed ceiling soffits throughout the building. We therefore took advantage of the raised floor needed to provide electrical distribution to the office wing and placed a series of upstand beams within it to stiffen the slab. This solution allowed the slab to be extremely thin and enabled the cantilever to be formed. The use of upstands was also a cost effective solution as it did not require complicated formwork and enabled the "table form" shuttering system, used elsewhere on the project, to be employed.

It was decided at the start of the design process to develop a base option for the structural frame and in parallel look at other frame options that could be cost effectively employed on this project. The base design, used in the employer's requirements, was a reinforced concrete flat slab and the various options developed included post-tensioning the slab and using precast units for both vertical and horizontal elements. The use of a steel frame had been ruled out due to the difficulty of detailing the exposed flat soffit for the ceiling. The reason for providing a base design and at the same time part-developing various options was to pre-empt the possible solutions that would come out of the procurement process. With most Design and Build projects the tendering contractor will offer up an alternative design in the hope that it will give them a competitive edge in terms of cost, quality or programme. It is difficult for a design team to predict what will come out of the procurement process although most contractors–and other members of the supply chain–will react to changes in material prices or other market conditions. So our role, as designers, must be to ensure that the solutions offered by the contractor can be effectively 'bolted on' to the developed design without expensive, time consuming or damaging changes to the design and architecture.

The tender returns for the Earth Sciences project had some contractors aligning their approach closely to the employer's requirements and others looking to bring in options such as post-tensioning and precast manufacture as we had predicted. As a result of the work undertaken in addition to the base option, we were therefore able to review the various frame options very quickly and identify the associated benefits and disadvantages. The chosen contractor, Laing O'Rourke, opted for post-tensioning throughout the scheme and the use of precast elements for all main cores, columns and shear walls, in order to reduce the slab thickness on the frame and speed up the construction process. The structural design was flexible enough to accommodate these in a number of ways. The design of the precast columns, for example, was based on a pinned top and bottom detail to enable them to be connected more easily. To cope with the increased compressive stresses, a single dowel reinforcing bar was designed to be used to assemble the precast columns on site, with conventional links introduced into the slab under the column's footprint to increase the compressive capacity of the concrete and thus counterbalance the increased stress on the dowels. It was agreed that the slab reduction achieved by using post-tensioned slabs would be taken up by an increase in the floor-to-ceiling height, and therefore no changes to the facade (and therefore to the planning requirements) were needed. Finally, the robustness of the precast cores introduced by Laing O'Rourke was considered. Additional ties could be accommodated within the landing slabs to retain the stiffness of the stability cores. The stability was checked to determine where thinner walls could be used, ensuring that full benefit was gained from the increased compressive strength of the precast elements in comparison to in-situ reinforced concrete. This process allowed the design team to choose the best contractor for the project with the most economical frame. It also prevented unforeseen difficulties and complexities from occurring which would erode the potential cost/programme saving.

A final challenge for the team was to ensure that no additional costs would be attributed to the build as a result of the University's aspiration to achieve an average of 40 per cent carbon reduction for all structural elements. At the Earth Sciences building the carbon reduction target was achieved by using Ground Granulated Blast Furnace Slag (or GGBS), a waste product from the steel industry with no carbon footprint. GGBS is used as a cement replacement in the concrete mix, and can cause a delay in the striking times of the shuttering system as the concrete takes longer to achieve the early strength gains required for it to support its own weight. The normal approach is to maximise the amount of GGBS used on non-critical elements or elements such as piles and ground beams that do not need their full strength immediately. The amount of GGBS used on the upper floors is then decided on a slab by slab basis taking into account weather, programme and follow-on trades. Laing O'Rourke's proposal to use post-tensioning allowed the project to achieve this aim without relying purely on cement replacement, as less concrete and steelwork would be required. This allowed the team to provide the required saving without incurring any cost of programme delays. Laing O'Rourke, however, still managed to use up to 20 per cent GGBS on the upper floors which added considerably to this building's green roots.

Appendices

Project team

Client	University of Oxford
Project Manager	RBDML/Oxford University Estates Directorate
Architect Landscape Architect	Wilkinson Eyre Architects
Building Services Engineer Fire Engineer Acoustics	Hoare Lea
Structural Engineer Traffic & Transportation	Pell Frischmann
Cost Manager	EC Harris
CDM Co-ordinator	HCD Management
Planning Consultant	Turnberry Consulting
Laboratory Planning	HDR CUH2A
Principal Contractor	Laing O'Rourke
Specialist M&E Subcontractor	Crown House Technologies

Wilkinson Eyre Team

Stafford Critchlow (Project Director), Ben Dawson, Deborah Dix, Jim Eyre, Anneli Giencke, Tony Musson, Rodrigo Piwonka, Glen Tomlin, Jan Vogel, Chris Wilkinson, Anna Woodeson, Sam Wright

Laing O'Rourke Team

Mel Banks, Wendy Barrow, Jon Bell, Dean Blackburn, Tony Brown, Frank Connolly, Steve Cooke, Owain Dobson, Mark Kane, John Kenneally, Chris Kerzel, Colleen Kneafsey, Phil Larkin, Jared Lazar, Mike Morris (Project Director), Rob Powell, Ajaz Syed, Dave Taylor, Joe Townsend, Chris Wiles

Hoare Lea Team

Peter Blick, Steve Buckley, Simon Dove, Ian Durbin (Project Partner), James Ford, Keith Horsley, Matt Jones, Andy Philips, Gael Vilatarsana, Mike Williams, David Woolley

Stone Sources

Jura Limestone	Neumeyer & Brigl, Eichstätt, Germany
Purbeck	Landers Quarries, Dorset, UK
Clipsham	Stamford Stone Quarry, Rutland, UK
Granite	Presa, Arcozelo, Portugal
Pietra Serena	Il Casone SpA, Firenzuola, Italy
Basaltite	Bolsena, Lazio, Italy

Subcontractors

Demolition	Keltbray
Excavation	Expanded Structures
Piling	Dew Piling
Ground source heating	Geothermal International
Concrete works	Expanded Structures
Steelwork	Shipleys
Cladding	M Price
Roofing	Prater
Screeding	Tellings
Flooring	AC Flooring
Stone flooring	WB Simpson
Joinery	Panel Edge Products & PMD
Doors & frames	Panel Edge Products
Partitions	C G Reynolds
Ceilings	SCS
Metalwork	Gascoyne & Beever
Raised access floor	Accsys
WC cubicles	Amwell
Lifts	Otis
Stone panelling	Malling
Laboratory fit-out	Waldner
Fume cupboards	PF&F
Field-free/metal-free labs	Felcon
Painting & decorating	F H Harvey
Blinds	Labetts
Scaffolding	H&H Scaffolding
Fire protection	Kilnbridge
External works	Vetter & C J Furey
Planting	University Parks
Signage	Rivermeade
Reception & library desk	Windsor Workshop

Suppliers

Stone	Vetter & Realstone
Paint	Dulux
Laminate	Formica
Dry wall	Knafe
Carpets & flooring	Interface
Ceilings	SAS
Lighting	Whitecroft
Doors	Panel Edge Products
Ironmongery	Bellsure
Glass	Pilkingtons
Glass cladding	Alsecco
Blinds	Labetts
Laboratory furniture	Waldner
Desks	Office Team
Seating	Auditoria
Shelving	Marstor
Signage	Rivermeade
Concrete	Cemex
Reinforcement	Roe
Twin wall	Oram
Precast columns	Malling
Sanitaryware	Armitage
Boilers	Broag
Chillers & air handlers	Trane
Ductwork	Ventair
Pumps	Wilo
Sockets & switches	MK
Granite	Hardscape

Facts and figures

Project data

Contract value	£29.5 million
Gross internal area	7,100 m^2
Cost of shell and core	£3,300/m^2
Cost including specialist laboratory fit-out	£4,100/m^2
Contract	Two-stage design & build
Design team appointed	September 2005
Started on site	July 2008
Date of occupation	October 2010

Facts about the department

Number of building users	296

Department of Earth Sciences established 1985 (previously the Department of Geology, dating back to the nineteenth century).

About 800 earth scientists have been trained since the department was established in its current form.

Construction statistics

Total concrete used	3,400 m^3
Reinforcements	400 t
Slab area (formwork)	7,000 m^2
Precast columns	154
Bulk excavation	5,800 m^3
Sheet piling	1,344 m^2
Luminaires	1,709
Sub-main cabling	6,200 m
Final circuit cabling	136,068 m
Sockets	2,400

Environmental performance

Treated floor area (i.e. where energy use can be measured) = 6,800 m^2

Annual energy/CO_2 consumption for space and water heating = 2,100,000kWhrs (= 58.7kgCO_2/m^2)

Annual energy/CO_2 consumption for electrical usage = 1,350,000kWhrs (= 85.4kgCO_2/m^2)

CO_2 emissions/m^2 treated floor area = 144.1kgCO_2/m^2

Ground source system

The system comprises 63 closed loop boreholes, underneath the footprint of the building and to a depth of approximately 65 metres, at an average spacing of approximately five metres. This creates a closed loop ground source energy system (GSES). Three reversible heat pumps in the basement plant room produce low temperature hot water at 45°C for heating and chilled water at 6°C for cooling.

The system is predicted to provide 1078 MWh of heating per year (61 per cent of the building's annual heating requirement) and 730 MWh of cooling (94 per cent of the building's annual cooling requirement).

It is also predicted to save 31 per cent of the building's delivered energy requirement and 17 per cent of its carbon emissions when compared to traditional building energy sources (i.e. gas-fired boilers and air-cooled chillers), with a 17 year payback period based on projected energy prices.

Other environmental information

Energy metering and sub-metering are linked to the building management system, allowing energy and water consumption to be monitored by the Building Manager.

Displacement ventilation is used for the internal offices with the return air path via the atrium and heat recovery at high level.

The teaching laboratories have mechanically-assisted natural ventilation (extract only).

Water from a 10,000 litre rainwater harvesting and stormwater attenuation tank is filtered and used to flush WCs throughout the building.

Drainage built into the external landscaping connects to an additional 200,000 litre stormwater attenuation tank.

Variable speed fans and pumps match air and water flow rates to demand.

Highly efficient light sources and lighting controls (including room occupancy sensors) have been specified throughout.

The chilled water system for the laboratories is designed as a closed loop to avoid 'once through' water usage and reduce waste.

The U-value (which defines the thermal transmittance performance of the building envelope) of selected areas is in excess of the minimum statutory requirements.

The building air-tightness criteria are also in excess of minimum statutory requirements.

External solar shading (via blinds or brise soleil) is used to effectively control solar gain to the naturally ventilated south and east-facing windows of the office wing and atrium.

The adoption of an in situ post-tensioned slab structural system reduced the total volume of concrete used by ten per cent.

25 per cent of the in situ cement used in the post-tensioned slabs was replaced with ground granulated blast-furnace slag (GGBS).

20 per cent of the in situ concrete infill to the precast concrete twin-wall elements was also replaced with GGBS.

38 per cent of the net internal area is naturally ventilated.

Exposed concrete soffits have been used to aid the night-time cooling of the spaces with natural ventilation or mechanically-assisted natural ventilation.

The more heavily serviced laboratory wing has thermally massive (stone-faced precast concrete) and highly insulated facades.

The sedum roof to the office wing provides biodiversity and rainwater attenuation.

All deleterious materials were eliminated from the construction process where possible, based on the Building Research Establishment's Green Guide to Specification.

All timber and timber products used are certified as originating from sustainably managed sources.

Contributors' biographies

John Hood

John Hood was Vice-Chancellor of the University of Oxford from 2004 to 2009. He is currently President and CEO of the Robertson Foundation in New York. He enjoys a range of other responsibilities including directorship of two British companies and a New Zealand company. He undertook engineering studies at the University of Auckland graduating with a PhD in structural engineering, before studying at Oxford for an M.Phil in Management Studies. Following 20 years in the business world, when he led international companies in construction, building industries and forest products, he returned to academia as Vice-Chancellor of the University of Auckland. During that period he chaired the board of a specialty geotechnical engineering firm. His interests include architecture, engineering, construction and conservation, as well as resource based industries, including oil and gas. They have fostered his respect for Earth Sciences and his strong commitment to their advancement.

Hugh Pearman

Hugh Pearman is the architecture and design critic of *The Sunday Times* newspaper, and editor of the *RIBA Journal*. He writes for many other publications in Britain and overseas, and contributes to radio and television.

His books include *Contemporary World Architecture*, published by Phaidon and *Airports: a century of architecture*, published by Laurence King.

Hugh has served on the architecture panel of the Arts Council, chaired the Art for Architecture initiative at the Royal Society of Arts, and acted as assessor on several architecture and design competitions.

He is an honorary fellow of the Royal Institute of British Architects. He helped to establish the annual Stirling Prize for architecture in 1996, and was a judge for its first three years.

His website www.hughpearman.com, otherwise known as Gabion, acts as an archive of his writing.

Hugh is married with four children, and lives in a house with 15 internal doors, all equipped with Arne Jacobsen door handles.

Philip England

Philip England is Professor of Geology in the University of Oxford. His undergraduate education was at the University of Bristol, where he read Physics. Although he took the first-year Geology course, there, his interest in Earth Sciences was aroused by lectures from Charles Frank in the Physics department, who showed that simple physical models can have great explanatory power in Geology. During his D.Phil research at Oxford, he investigated the Earth's heat flow and the origins of metamorphic rocks. (He was happy to find a practical application of this knowledge when specifying the ground-source heat pump for the building.) He subsequently worked at Cambridge and Harvard, before returning to Oxford in 1986. His research is driven by the belief that the complexities of the natural world often hide an underlying simplicity. The search for that simplicity has led him to investigate several aspects of tectonics, including the growth and decay of mountain ranges, and the origins of the arc volcanoes –the "Ring of Fire". He is currently obsessed by the earthquake hazard in the developing world.

Stephen Hesselbo

Stephen Hesselbo is Professor of Stratigraphy in the Department of Earth Sciences, University of Oxford. His main research interests are in major environmental changes through Earth history, particularly for that interval of the Mesozoic that is known to geologists as a "greenhouse world". Educated in Scottish geology where he grew up in North Berwick, and as an undergraduate at the University of Aberdeen, he completed a PhD at the University of Bristol. He is a field geologist at heart, and has many years of experience undertaking research expeditions in regions as varied as Greenland, NW China, and Argentina; at the same time he has a special interest in the Jurassic geology of the British Isles, including sedimentary rocks traditionally used as building stones. He also has wide experience of the international Ocean Drilling Program, a long-running multinational research effort based on recovery of rock cores from the sea floor. In 2004 he was awarded the first Medal of the International Commission on Stratigraphy to honour high quality research in stratigraphy.

Chris Wilkinson

Chris Wilkinson is a principal and founder of Wilkinson Eyre Architects. After working in a number of top UK practices, he set up Chris Wilkinson Architects in 1983, formed a partnership with Jim Eyre in 1987 and established Wilkinson Eyre in 1999. With a passion for architecture and design, he has helped develop the profile of Wilkinson Eyre Architects as a leading international architectural practice. In addition to his management coordination role, he leads design and takes responsibility for projects across a broad range of sectors with his co-directors. Alongside the Earth Sciences building at Oxford, he has been closely involved with many of the practice's recent buildings in the education sector, including the Forum Project at the University of Exeter, the new Humanities department at Queen Mary, University of London, the John Madjeski Academy and Bristol BSF Schools.

Chris combines a lifelong interest in art with a fascination for science, technological innovation and a sense of history, producing a fresh new approach to architecture. The design for every project is different, and relates closely to the specificity of brief and context. His contribution to architecture has been recognised by the award of an OBE in the Millennium Honours List, election to the Royal Academy of Arts in 2006, and an Honorary Fellowship of the American Institute of Architecture in 2007. Chris was appointed as Architectural Commissioner of English Heritage in 2006, made Patron of the Civic Trust Awards in 2009, and has been awarded Honorary Doctorates by Westminster University and Oxford Brookes University. In addition, he was voted Designer of the Year by CSD and the Design Council in 1996 and FX Designer of the Year in 2001.

Stafford Critchlow

Stafford Critchlow is a Director at Wilkinson Eyre Architects. He studied architecture at Newcastle University and for a year in Venice at the Istituto Universitario di Architettura Venezia (IUAV). Prior to joining Chris Wilkinson nearly 20 years ago he worked for Skidmore Owings & Merrill, Gaul Associates in Chicago and at the Boilerhouse Project in the V&A Museum. Stafford has been responsible for many of the practice's buildings for education starting with Explore@Bristol and two award-winning projects for City and Islington College in London. In 2003 he led the team designing an 'exemplar' Secondary School for the Future for DfES: many of the pedagogical ideas from which were realised at the John Madejski Academy in Reading and the first BSF schools in Bristol. He has Higher Education projects on site or completed at Exeter University, Queen Mary University of London and at Swinburne in Melbourne, Australia.

In 2005 his work was chosen for the *Architects Journal* 40 under 40 group show at the V&A Museum. He is interested in architecture as a vehicle for change, both social and in the academic sphere, and in art in architecture. He is married to the writer Harriet Lane, has two children and built a house extension at the same time as the Earth Sciences project was on site: as a tribute he used the same Pietra Serena stone in his back yard.

Ian Durbin

Ian Durbin is a Partner with more than 25 years experience with Hoare Lea. Ian graduated from Bath University's pioneering School of Architecture and Building Engineering course, where he was introduced to multi-disciplinary working and the holistic approach to creating new buildings.

Ian has a strong track record and specialism in science and research facilities and has worked extensively in Oxford and Cambridge. He was responsible for leading the Hoare Lea team, which included lighting, acoustics and fire engineering expertise, for the Earth Sciences project.

Ian was responsible for the establishment of Hoare Lea's Oxford office in 2008; this is now a thriving design-led studio, which works closely with Oxford University and others in the education, science and heritage sectors.

Ian has a passion for simple, straightforward design solutions and is championing the use of low carbon technologies and design techniques, particularly in science and research buildings.

Keith Horsley

Keith Horsley is an associate with Hoare Lea. He has 20 years' experience in the industry and has led the design of the mechanical and electrical engineering services on a variety of projects in the commercial, education, residential, leisure and laboratory sectors.

Keith now heads the group within Hoare Lea that is responsible for knowledge management and technical control and a large part of his role involves the development and internal dissemination of Hoare Lea's knowledge and experience of low and zero carbon energy systems.

Keith was lead mechanical engineer on the Earth Sciences project during the briefing, outline design and scheme design stages, and led the design and energy analysis of the ground source energy system throughout the project. He has also been involved in several other Hoare Lea projects in which ground source energy has been used, including One New Change, the mixed-use development in London.

Mike Wigg

Mike is Head of Capital Projects within the Estates Directorate of the University of Oxford, a post he has held since 2003. His team is responsible for the delivery of all major new build and refurbishment projects for the University, with an annual capital spend currently of circa £100 million.

Mike initially trained as a civil and structural engineer, working on the design of major highway structures and a range of new buildings. After a short time running the in-house design group for a design and build contractor, he moved to head up the "New Works" team for the United Kingdom Atomic Energy Authority based at Harwell in Oxfordshire. Following this, and prior to moving to Oxford University, he was based in London where he worked for 14 years in commercial development, for the majority of this time heading the project management team for MEPC.

Mike Morris

Mike joined John Laing Construction in 1974 as an articled pupil in the Quantity Surveying department in Birmingham. He worked on numerous schemes in the Midlands during the 1970s and 1980s including Walsgrave Hospital, Coventry Building Society and Birmingham Airport. In 1988 Mike became project surveyor on the Raven Meadows shopping centre in Shrewsbury, and then carried out similar roles on PowerGen's headquarters in Coventry, the Britannic Assurance headquarters and MEPC's office development in Colmore Row in central Birmingham.

In 2001 Mike came down to Oxford University for an initial six week period to help finalise the bid for the Chemistry Research Building. Three months later, when the price was agreed with the University, Mike was asked to take up the challenge and become Project Manager for the job. This he did and the job was a great success, earning him the CIOB Construction Manager of the Year award in 2004. Mike then returned to Birmingham for two years, leading the team for the University of Central England's new facility in Edgbaston. He returned to Oxford in 2006 as Project Director for the new Biochemistry Building, for which he was again awarded Construction Manager of the Year for 2009. In 2008 the Laing O'Rourke team, under Mike's leadership, were appointed as main contractor for the Oxford Earth Sciences project and the project was successfully completed in September 2010.

David Sivyer

David Sivyer was the Technical Director at Pell Frischmann responsible for the structural design of the Earth Sciences Building. He has over 15 years of consulting experience since graduating in 1996. The son of a contractor, David has always had a passion for building and a keen appreciation for on site collaboration. In 2000, David became a Chartered Member of the Institution of Structural Engineers, which in the same year presented him with the Graham Wood award.

David has led many interesting projects across varied building typologies, from the construction of the new SAS base in Hereford to the refurbishment of a Burberry store in Barcelona. Recently several of his projects have been recognised by the Association for Consultancy and Engineering (ACE) in their annual Engineering Excellence Awards scheme–in 2009 with 7–10 Old Bailey, and in 2011 with the Earth Sciences project itself.

Since Earth Sciences, David has gone on to lead the structural design of many more buildings in Oxford, including the Humanities Centre, the New Bodleian Library refurbishment, the new Mathematical Institute and Chemistry Research Laboratory II. He is currently embarking on a new challenge as the Director of Structures for the London office of Blyth and Blyth.

Wilkinson Eyre Architects

Wilkinson Eyre Architects is one of the UK's leading architectural practices, with a portfolio of national and international projects. Our architecture is based on an informed use of technology and materials and combines a commitment to the spirit of the new with a strong awareness of context. This approach to design is based on the establishment of a clear brief and a legible working diagram through close liaison with the client from the earliest stages of a project. The practice's reputation is founded on a commitment to quality, programme and value for money which is demonstrated by a number of repeat commissions for leading clients. The designs of Wilkinson Eyre Architects are widely recognised and have received extensive media, public and professional acclaim and numerous awards including, uniquely, the RIBA Stirling Prize in two consecutive years.

The practice has designed highly successful projects across a wide range of market sectors, and has considerable experience in designing buildings for education, ranging from 'schools for the future' through to highly specialist university research facilities. Our work with higher education clients began with a series of projects for Anglia Ruskin University, including the award-winning Ashcroft Business School. In 2006 we were appointed by the University of Oxford to design the new building for the Department of Earth Sciences–and subsequently the refurbishment of the New Bodleian Library. We have designed a series of projects at Queen Mary, University of London, and in 2010 completed the Hauser Forum at the University of Cambridge. Key current commissions are the Forum Project at the University of Exeter and an Advanced Manufacturing Centre for Swinburne University of Technology in Melbourne, Australia. Closely linked to these education projects is a extensive portfolio of research-based buildings, including an R&D facility for Dyson in the 1990s and the Jodrell Laboratory at the Royal Botanic Gardens at Kew.

Anglia Ruskin University
Ashcroft International Business School/ Rivermead Campus

The commission to design the Ashcroft International Business School at Anglia Ruskin University followed an international design competition, the brief demanding a building of architectural distinction to house a centre for learning, teaching and research in disciplines related to management and business. The award-winning building was acclaimed for its energy-efficient design, and Wilkinson Eyre were subsequently commissioned to masterplan the University's new Rivermead Campus near Chelmsford. The new campus has been set out in sweeping arcs which respond to the curve of the River Chelmer. The design actively promotes high environmental performance in new buildings on the campus, which will be enhanced by the design of the landscape and external spaces. The first phases of the masterplan have now been implemented by us, starting with a Student Centre and multi-purpose Sports Centre. A new Health School has also been completed, fronting onto a landscaped square at the heart of the campus adjacent to the river.

Ashcroft International Business School completed 2002; Student Centre/ Sports Centre 2006; Health School 2007

Royal Botanic Gardens Kew
Jodrell Laboratory

Wilkinson Eyre's extension to the Jodrell Laboratory at Kew was designed to improve the current research facilities and provide sufficient space to bring together a number of related scientific specialisms. The project acknowledges the framework of the Site Development Plan (prepared in 2002 by Wilkinson Eyre) and meets Kew's objectives of improving accessibility and the public's awareness of science at the gardens. The principal design strategy was to extend the existing laboratory to create a new three-sided courtyard space in conjunction with the School of Horticulture building to the south, giving a collegiate feel to the complex which is now focused around the Aquatic Garden. The extension provides 2000 square metres of accommodation split between herbarium storage, a library, laboratories and office space. The existing building has been further extended by infilling the colonnade of an earlier extension. The redevelopment provides a new face to the laboratory, effectively turning the building around to establish a stronger relationship between members of the public visiting Kew and the research activity within.

Completed June 2006

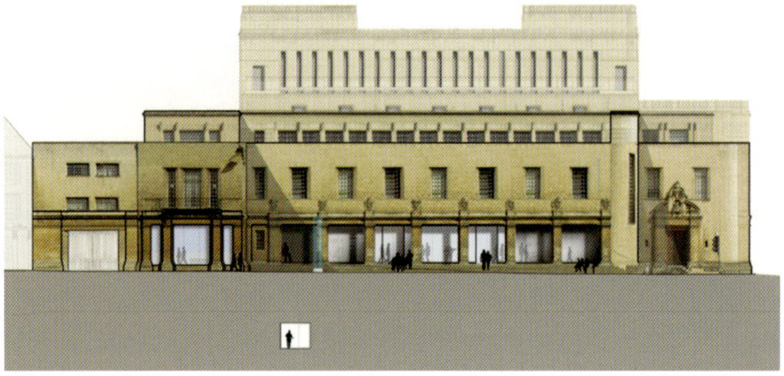

University of Cambridge
Hauser Forum

The Hauser Forum, named after the building's alumni sponsor, forms a new gateway to the University's West Cambridge Campus. This new site on greenbelt land to the west of the city houses a number of forward-thinking departments including Nanotechnology and Physics for Medicine. The Forum provides office and research facilities for small start-up companies developing spin off technologies from other University scientific research programmes. It consists of three elements–an incubator building, a R&D building and a large cafe space. The cafe encourages the exchange of ideas between academics and visitors, and the image of the Forum as a place to meet and exchange ideas is reinforced by an oversailing canopy which unites the building elements and shelters a public plaza below. The development is naturally ventilated and was designed to achieve a BREEAM Excellent rating. It incorporates many environmental features including a labyrinth below the building to pre-cool incoming air, ground source heat pumps and an attenuation pond and reed beds to filter run-off water.

Completed May 2010

University of Oxford
New Bodleian Library

At the heart of Oxford's historic core, Gilbert Scott's 1940 New Bodleian Library is a vital resource for academic research. In 2006 Wilkinson Eyre was appointed to refurbish the library as a new cultural and intellectual landmark. The essence of our concept is an inversion of the traditional Oxford quadrangle, playing with contrasts of light and darkness, void and volume. The bookstack itself becomes a simple yet dramatic box, poised above a generous public area which permeates the building. The design works with, rather than against, Scott's robust design, reinvigorating the space with an improved circulation diagram–for both book retrieval and user movement–and creating a number of contemporary interventions, including a spectacular reading room at roof level. This, together with a series of additional reading rooms, will reinforce the library as a major new research resource for special collections. Careful consideration has been made to the library's position within the surrounding cityscape, opening up new ground level entrances to encourage public access and help knit the building into the wider urban grain.

Appointed November 2006

Queen Mary, University of London Humanities Building/Mathematical Sciences Building

Queen Mary, University of London, is ranked fourth in the UK for the provision of arts and humanities courses. This project forms part of the college's ongoing development programme, providing a new humanities building to enhance the existing campus on the Mile End Road. The building accommodates the entire History department and provides additional learning space that is centrally bookable for other departments. Located on the main road, it provides the School of Humanities with an important new public face on a major arterial route between the City of London and the 2012 Olympic Park. It also incorporates a variety of innovative sustainability measures, including ground source heat pumps for the generation of renewable energy. Wilkinson Eyre were subsequently commissioned to create a new entrance lobby and social/study space for the Mathematics department, again establishing a strong identity for the client on the Mile End Road, with a single-storey structure clad in screen printed glass tiles. The design uses the non-repeating 'Penrose' pattern of kites and darts, sanctioned for use here by Sir Roger Penrose, a professor in the department.

Completed Spring 2011

University of Exeter Forum Project

Following an international design competition, Wilkinson Eyre was appointed in 2008 to design this new centrepiece for the University of Exeter's famously hilly Streatham Campus. Working with the natural features of the site, the scheme creates a 'green corridor' to connect the Forum with the wider landscape. Central to the scheme is an undulating timber gridshell roof, which shelters and unifies a series of new student-focused spaces within. The fluid form contrasts with the orthogonal brick volumes of the existing buildings on this steeply sloping site, and respects key views across the city to Dartmoor. The Forum also features an extended and refurbished Library, new learning spaces, student services, catering and retail outlets, a landscaped plaza and new University reception, and forms part of a £450 million capital investment programme which has helped propel Exeter into the top ten UK universities.

Appointed October 2008

Laing O'Rourke

Laing O'Rourke is the UK's largest privately owned construction solutions provider, with the Group's international operations posting managed revenue of £4.3 billion in 2010. The business is made up of a range of engineering, construction and specialist services companies, which combine to provide prestigious clients with a comprehensive investment, development and management capability.

Working across Europe, the Middle East, South Asia and Australasia, our operations span eight core sectors: lifestyle; business; social infrastructure; transport; power; mining and natural resources; oil and gas and utilities and waste. The breadth of our expertise and geographical reach means we are engaged on the full spectrum of projects–from remote desert railways to urban regeneration.

At the forefront of bringing manufacturing and offsite construction methods to bear in the construction industry, our offsite manufacturing capabilities enable faster construction and greater standardisation of core components–essential to meet the demands of this growing sector and ensure the durability of heavily used facilities.

We are a leading supplier of social infrastructure for governments and private-sector clients, with a particularly strong presence in the field of higher education. The Earth Sciences building is just one of a series of projects undertaken by Laing O'Rourke for the University of Oxford, including the Chemistry Research Laboratory and the Biochemistry Building. The company recently reinforced its commitment to the University by collaborating with the Department of Engineering Science to establish a new centre for engineering focusing on interdisciplinary research relevant to the construction industry.

Queen Mary, University of London Institute of Cell and Molecular Science

This two-stage design and construct project used the combined strengths of the Laing O'Rourke Group to produce a world-class medical school at the cutting edge of teaching and research. The 9,000 square metres Blizzard Building incorporates a large laboratory providing accommodation for up to 395 researchers. This receives natural light from a pavilion and rooflights. The glazed pavilion houses open-plan desks, cellular offices and suspended seminar "pods", and a public exhibition space called the "Centre of the Cell", designed to encourage children to develop an interest in science. A bridge link provides connections to the existing Clinical Sciences Research Centre. The laboratory spaces are considered to be Queen Mary's best-performing Cat 3 labs, with air leakage rates at least 25 times better than required and subsequently adopted for the next laboratory building. In 2006 the project received awards from the RIBA and the Civic Trust.

Completed March 2005

NHS Blood and Transplant Service National Blood Centre, Filton, Bristol

Laing O'Rourke was appointed as Preferred Supply Chain Partner under the Procure 21 Initiative for this new centre which is the South West's principal blood service facility. Designed by BDP, it replaces existing outdated accommodation and provides the necessary technical, environmental and administrative facilities to process, test, validate and package donations of human blood. The development is the first of three new UK Blood Centres projects. This facility has a gross internal floor area of 14,000 square metres, and mainly consists of laboratory spaces environmentally controlled to a CL2 level of cleanliness. The centre includes manufacturing facilities to allow the blood to be processed as a clinical product; this area has a higher level of particulate control than the general laboratories. It also houses administrative accommodation to support the technical areas, communal facilities such as seminar rooms, teaching facilities, library, restaurant, rest areas, changing rooms and toilets.

Completed June 2008

University of Oxford
Biochemistry Phase 1

Laing O'Rourke was the principal contractor on a project that was designed to facilitate and encourage collaboration between researchers and research groups. The 12,000 square metres building is intended to 'wow' users and challenge preconceptions of how laboratory buildings should be designed, marking the first phase of redevelopment in the University of Oxford's Science Area. Designed by Hawkins\Brown, Biochemistry Phase 1 has enabled the department to collocate all of its main research activity into a single building. To satisfy the Carfax height planning constraint, the two storeys of accommodation are below ground with the other two above ground. The project included cycle parking and new pedestrian areas to reinforce the overall Science Area master plan.

Site restrictions dictated two phases of building construction, with significant numbers of staff decanted from their existing accommodation to allow demolition while maintaining research and teaching activities.

Completed July 2008

University of Oxford
Chemistry Research Laboratory

Oxford University is at the forefront of Chemistry research, having produced four Nobel Prize winners. Laing O'Rourke, as principal contractor, delivered this flagship research facility on a two-stage contract. This new laboratory, designed by RMJM, reaffirms the University's position as a pioneer. Located in the heart of the Science Area, it provides 16,500 square metres of space, including a two-level top down basement (a specialist construction skill of Laing O'Rourke) and 48 laboratories and secondary areas linked to a three-storey office block adjacent to a glazed atrium. The laboratories house a range of cutting-edge instruments, including magnetic lasers, nuclear magnetic resonance machines and mass spectrometers–with protection provided by a steel shield built into the basement wall.

Project Manager, Mike Morris, won the 'New Build Between £15 million and £45 million Category' Gold Award at the CIOB Construction Manager of the Year Awards, 2004.

Completed March 2003

Maidstone and Tunbridge Wells NHS Trust Pembury Hospital PFI

Laing O'Rourke is design and construct contractor for this project, the first major new hospital within the UK to comprise 100 per cent single bedrooms with en-suite bathrooms. The scheme, designed by Anshen+Allen, has evidence-based design at its core and includes micro-planning of patient bedrooms and ward areas to improve patient safety and maximise staff efficiency.

The new 66,000 square metres hospital will be essentially a 'hot' site, with 513 predominantly emergency beds and a relatively small number of elective beds. The scheme also includes a women's and children's hospital, emergency care centre and outpatient facilities. The latter responds to the changing approach within the NHS, with emphasis on moving services out of acute settings, where possible. Circulation patterns within the hospital keep different staff, patient and visitors' groups separate, which, in turn, enhances the whole hospital environment, improves efficiency and makes a positive contribution to wellbeing.

Phase 1 completed May 2011

University of Cambridge Cancer Research UK Cambridge Research Institute

This new-build laboratory was designed to house 30 research groups of ten to 15 people–the majority funded by the Imperial Cancer Research Fund; the others a mix of University and University/Cancer Research UK joint groups. The project involved the construction of a five-storey, 14,000 square metres Cancer Research Centre with basement, offices, lecture theatre suite, exhibition area, energy centre, ancillary buildings, restaurant, car parking and associated external works and services. Works followed the completion of the adjacent multi-storey car park.

The structure comprises a reinforced in situ concrete frame, with structural steelwork to the plantroom and ancillary buildings. The external envelope consists of glazed curtain walling, metal cladding, and brickwork and blockwork. Internally, drylined partitions are used, with traditional finishes. The contract also included installation of laboratory furniture and a CAT 3 suite of modular construction, while the third floor was left as a shell for future development.

Completed September 2006

Hoare Lea

Hoare Lea is an award-winning firm of consulting engineers specialising in mechanical, electrical and public health (MEP) engineering. Founded in 1862, Hoare Lea remains an independent partnership with a national network of 12 offices and a growing international presence. We lead the way in sustainability and our reputation for an innovative approach, backed up with strong delivery, underpins our ability to create long-standing relationships with our clients. We work on some of the highest profile projects in the industry and with leading architects, designers and contractors, and are currently working on projects across all main market sectors, ranging from small studies and reports to projects valued at more than £4 billion.

We have extensive experience in the education, science and research sectors and have worked with many universities, including Cambridge University, Imperial College, and University College London. At Oxford University, in addition to the Earth Sciences building and other projects, Hoare Lea is working on the new Chemistry Oxford research laboratory, which aims to be the "greenest Chemistry building in the world".

Our engineering excellence has contributed to many award-winning schemes, such as Horizon House, Bristol, winner of the BREEAM Offices Award 2010; and One New Change, London, Overall and Mixed Use Winner of the MIPIM Architectural Review Future Project Awards 2010 and winner of a regional RIBA (Royal Institute of British Architects) award in 2011. In 2010 Hoare Lea won both 'Infrastructure' and 'Building Services' categories at the Association for Consultancy and Engineering (ACE) Engineering Excellence Awards for its work at Cabot Circus, Bristol, and Warwick Digital Laboratory. In 2011 the practice won the Integrated Project Team Award at the CIBSE Building Performance Awards for Ceredigion County Council's new office building, Canolfan Rheidol, in Aberystwyth.

Lancaster University
Lancaster Institute for the Contemporary Arts (LICA)

Lancaster University has a strong commitment to sustainability throughout its campus and Hoare Lea has worked with the University to achieve award-winning, BREEAM Excellent, sustainable buildings such as InfoLab21 and the Postgraduate Statistics Centre. LICA was the next step on this journey. The three storey 5,000 square metres LICA building is an exciting purpose-built facility which combines teaching and research activities in Art, Design, Film Studies, Music and Theatre Studies across a range of interactive and flexible spaces.

A key part of designing LICA was to concentrate first on passive design to reduce energy demand. Hoare Lea collaborated with architects Sheppard Robson, testing a range of materials, and the choice of a pre-fabricated all-timber construction achieved the lowest affordable U-value solution and a very low air infiltration rate. As both building services engineer and BREEAM Assessor, Hoare Lea worked with the client and project team to achieve a BREEAM score of >80 per cent. The design also incorporated roof-mounted photovoltaics and rainwater harvesting, both with information displays; displacement ventilation; 75 per cent efficient thermal wheel heat recovery; and a link to the campus heating.

Completed September 2010

Imperial College London
L Block Development

The L Block development on Imperial College's Hammersmith campus will provide a 13,000 square metres, six storey scientific research facility for Imperial's Academic Health Science Centre, including one of the largest cardiovascular research facilities in Europe. Hoare Lea has provided mechanical and electrical design services and specialist advice including fire engineering, acoustics, thermal modelling, vertical transportation, facade access and BREEAM assessment.

The L Block, by architects Sheppard Robson, has been designed with challenging sustainability objectives, including a BREEAM Excellent rating and a carbon emissions performance 32.6 per cent lower than required by Part L 2006. Various features maximise energy efficiency, for example passive exposed chilled beams use open loop boreholes to provide cooling without the use of chillers. Variable volume fume-cupboards will reduce the heating and cooling of the associated make-up air and a site-wide 2MWe CHP-supported heating scheme will reduce carbon emissions. The redevelopment, including the demolition of the existing building, has been undertaken in a live hospital environment with access and site-wide services maintained.

Appointed September 2006

Land Securities
One New Change, London

Designed by Jean Nouvel with executive architects Sidell Gibson, the City of London's newest shopping and leisure destination One New Change covers 51,000 square metres over eight floors, with 60 shops and restaurants and high-grade offices. Hoare Lea designed the office and retail MEP systems for the Land Securities' development and provided specialist advice regarding vertical transportation, together with sustainability and energy systems expertise.

The building services design has provided an innovative and energy efficient solution. It features one of the most sophisticated commercial ground source heat pump systems in Europe, comprising 196 active energy piles and one pair of open loop boreholes. Thirteen reversible ground source heat pumps are arranged to allow flexibility to match the differing load profiles of the office and retail areas and provide heat recovery. The energy system is estimated to reduce carbon emissions by at least ten per cent and could save £300,000 on energy bills a year.

Completed October 2010

The Environment Agency
Horizon House, Bristol

The Environment Agency's national office, the 6,595 square metres Horizon House in Bristol, is one of the greenest office buildings in the UK. It won the 2010 Best BREEAM Office Award with a rating of 85.06 per cent–the highest awarded to an office building under the 2006 criteria. Hoare Lea acted as MEP consulting engineers from stage D onwards and provided specialist acoustic, fire and virtual engineering services.

The building, designed by Alec French Architects for developer Westmark, represents the coming together of several technologies into a single low-energy design. Ground source heat pumps, with 25 boreholes drilled 55 metres deep, assist heating and cooling, and with solar panels and photovoltaics meet 17.5 per cent of the building's energy needs. The result is a commercially viable building, built to the highest environmental standards, within a city centre location. Completed on time and on budget, the Environment Agency will make operational savings and productivity efficiencies of around £1.12 million per year.

Completed December 2010

University of Cambridge
Department of Materials Science and Metallurgy

This £48 million research building, designed by a team including architects NBBJ, structural engineers Ramboll, and MEP consultants Hoare Lea, is located on the West Cambridge Site and will provide a new home for the Department of Materials Science and Metallurgy.

The 10,700 square metres facility will accommodate a broad range of research fields from large scale processing of metals through to nanotechnology, Chemistry and the development of medical materials. The new purpose-built facility will contain a mix of laboratories, support facilities, offices and social space. It will include a world-class electron microscope suite built on a two-metre deep concrete slab to eliminate vibration and electromagnetic interference from external sources.

Hoare Lea's role encompasses MEP, lighting, fire engineering, sustainability and vertical transportation advice.

Appointed February 2009

University of Oxford
Chemistry Oxford

This new research and teaching facility for the Department of Chemistry at the University of Oxford will provide more than 20,000 square metres of new floor space and, with the existing Chemistry Research Laboratory, will create a world-class facility to be named "Chemistry Oxford".

Located in the University's Science Area, Chemistry Oxford will have three basement levels and contain more than 50 research laboratories as well as teaching laboratories, lecture theatres, seminar spaces, workshops and offices together with a cafe and integrated landscape design. The new building will be connected to the existing building below ground by a tunnel and above ground by a shared surface across South Parks and Mansfield Roads that will encourage links between the two.

Designed by FJMT and aiming to be the "greenest Chemistry building in the world", the building has strong environmental credentials and is targeting a BREEAM 'Excellent' rating with an aspiration towards 'Outstanding'. The design includes variable air volume low velocity fume hoods incorporating heat recovery, a very large ground source energy system, low energy flexible lighting and an innovative double skin ventilated facade.

Hoare Lea has provided mechanical, electrical and public health services design, together with specialist acoustics, fire and sustainability advice and BREEAM assessment.

Appointed July 2010

Credits

cover, 16–18, 21, 25–27, 29, 32–48, 49 (left), 50–51, 53–55 (stone details), 67–68, 72 (bottom), 74–75, 99, 119	Morley von Sternberg
9, 11–15, 20, 22–23, 57, 60–65, 66 (left), 73, 81	Edmund Sumner
138 (left)	Ben Blossom
129 (right)	James Brittain
31 (bottom), 69 (left)	Stafford Critchlow
28 (thin sections), 30 (left), 112-113	Department of Earth Sciences, University of Oxford
138 (right)	Simon Doling
31 (top)	Philip England
101	Estates Directorate, University of Oxford
102 (bottom)	Farrells
139 (right)	FJMT
102 (top)	Grant Associates/Farrells
104-105	Karl Harrison
30 (right)	Gideon Henderson
49 (right), 52, 53–55 (thin sections)	Stephen Hesselbo
95–97	Hoare Lea
137 (left)	Hufton+Crow
107–111, 114, 115 (bottom), 117, 133–135	Laing O'Rourke
80	Malling
130 (left)	Mott MacDonald
28 (left)	NASA, ESA, and The Hubble Heritage Team STScI/AURA
139 (left)	NBBJ
115 (top)	Pell Frischmann
137 (right)	Sheppard Robson
128 (left)	Nick Wood

Acknowledgements

Many people contribute to the success of major institutional building projects. Architects, planners, engineers, contractors, regulators, legal and other advisors, City personnel, University administrators and estates' professionals, all deserve recognition and thanks for their work on this project. Many members of the department devoted their efforts to ensuring the successful outcome of the project, and all of us are grateful to Mike Wigg, who set the project up in a way that allowed the department to participate effectively in the design of the building, to Chris Wilkinson and his team, who were responsive (but not too responsive) to these amateur interventions, and to Mike Morris, who lead an outstanding team of contractors. We are also deeply grateful for the cornerstone donation made by the Gareth and Gwenna Roberts Foundation, and for a key donation by the Wolfson Foundation, which supports the new Geochemistry Laboratories. These significant gifts enabled the project's timely realisation.

Professor Philip England

That this book has been written in collaboration with the Earth Sciences and Estates departments in Oxford and the consultant and contractor team is indicative of the spirit with which this project has been prosecuted from the outset. In supporting the idea of this book unequivocally from the very first discussion about the possibility I must thank especially Mike Wigg, Chris Wilkinson, Philip England, Ian Durbin and Mike Morris for their enthusiasm and unswerving support. In helping to produce this book we are especially grateful to Tony Musson, Anneli Giencke, Tim Thatcher and Bethany Wren at Wilkinson Eyre, to Hannah Jackson in the Department of Earth Sciences for facilitating the interviews and for the interviewees in giving freely of their time and opinions. Thanks also to Duncan McCorquodale, Libby Waite and Rachel Pfleger at Black Dog Publishing for taking us on, to Edmund Sumner and Morley von Sternberg for some sublime photographs and to Harriet Lane for casting a critical eye. Finally, none of this would have been possible without Emma Keyte to bring it all together. Thanks to her not just for her good humour in providing the editing and general coaxing skills required but also for her invaluable insights in bringing together the worlds of architecture, Earth Sciences, construction engineering and publishing.

Stafford Critchlow

Black Dog Publishing Limited
10A Acton Street
London
WC1X 9NG

t. +44 (0)207 713 5097
f. +44 (0)207 713 8682
e. info@blackdogonline.com

Project manager: Emma Keyte.
Designer: Rachel Pfleger at Black Dog Publishing.

British Library Cataloguing-in-Publication Data.
A CIP record for this book is available from the British Library.

ISBN 978 1 907317 53 8

Black Dog Publishing is an environmentally responsible company. *Tectonics: a building for Earth Sciences at Oxford* is printed on FSC accredited paper.

Printed in Iceland.